Annual Conference of the American Forage and Grassland Council (AFGC 2022)

Wichita, Kansas, USA
7-12 January 2022

ISBN: 978-1-7138-5976-5

Printed from e-media with permission by:

Curran Associates, Inc.
57 Morehouse Lane
Red Hook, NY 12571

Some format issues inherent in the e-media version may also appear in this print version.

Printed by Curran Associates, Inc. (2022)

For permission requests, please contact American Forage and Grassland Council (AFGC) at the address below.

American Forage and Grassland Council (AFGC)
PO Box 867
Berea, KY 40403
USA

Phone: 800.944.2342

www.afgc.org

Additional copies of this publication are available from:

Curran Associates, Inc.
57 Morehouse Lane
Red Hook, NY 12571 USA
Phone: 845-758-0400
Fax: 845-758-2633
Email: curran@proceedings.com
Web: www.proceedings.com

TABLE OF CONTENTS

INCORPORATING PERENNIAL FORAGES INTO SOUTHERN ROW CROPPING SYSTEMS 1
E. Billman, B. Campbell, F. Reay-Jones, G. Sigua

USING TECHNOLOGY TO MINIMIZE LABOR AND IMPROVE LIVESTOCK PRODUCTION UNDER GRAZING 2
C. Bagley, R. Palfreyman, J. Whitson

BIOCHARS IMPACTS ON PLANT GERMINATION AND GROWTH RATES WHEN LAND-APPLIED AND RUMINAL FERMENTATION RATES WHEN ADDED TO DIETS 3
C. Bagley, K. Han, R. Palfreyman, J. Whitson

USING SOCIAL MEDIA FOR FORAGE EDUCATION AND PROMOTION 4
L. Baxter, J. Tucker

IMPACT OF MANAGEMENT ON BERMUDAGRASS STEM MAGGOT DAMAGE 5
L. Baxter, B. Anderson, W. Hudson

PATCH-BURN AND ROTATIONAL GRAZING EFFECTS ON PERFORMANCE OF BEEF HEIFERS AND PASTURE PRODUCTIVITY 6
M. Berry, P. Keyser, R. Smith, E. Bisangwa

EXTRACTION OF LIGNIN MONOMERS FOR UPLC ANALYSIS 7
M. Boss, J. Strickland, M. Aguerre

GROWTH AND ENSILING OF REDUCED LIGNIN ALFALFA IN THE SOUTHEASTERN US 8
M. Boss, M. Aguerre, J. Strickland

MEASURING IMPACT OF FERTILIZATION ON BROOMSEDGE ENCROACHMENT IN HAYFIELDS USING REMOTE SENSING 9
J. Bush, C. Teutsch, J. Henning, S. Smith

EVALUATION OF REMOTE SENSING TECHNIQUES TO MEASURE CRABGRASS ENCROACHMENT IN TALL FESCUE STANDS 10
J. Bush, C. Teutsch, J. Henning, S. Smith

MEASURING PROPORTION OF LIVE AND DEAD VEGETATION IN PASTURES USING REMOTE SENSING 11
J. Bush, C. Teutsch, J. Henning, S. Smith

FORAGE PRODUCTION AND NUTRITIONAL QUALITY OF TALL FESCUE AND ALFALFA UNDER LONG TERM WARMING AND SEVERE DROUGHT 12
K. Catunda, A. Churchill, S. Power, B. Moore

EXTREME SUMMER DROUGHT IMPACTS ON OLD-FIELD FORAGE PRODUCTION AND NUTRITIONAL QUALITY 13
A. Churchill, F. Isbell

IMPACT OF BMR TRAIT ON DECOMPOSITION RATE OF SORGHUM-SUDANGRASS 14
B. Dooley, J. Bush, C. Teutsch, B. Hendrix

INVASIVE SPECIES ON KANSAS GRASSLANDS 15
W. Fick

CHLOROPHYLL FLUORESCENCE AS AN INDICATOR OF PLANT FORAGE STRESS 16
S. Goslee

COMPAIRISON OF FIVE METHODS OF BOTANICAL ANALYSIS ACROSS KENTUCKY 17
E. Gotsick, S. Smith, J. Henning, C. Teutsch

EFFECT OF AN IMPROVED GRAZING MANAGEMENT SYSTEM ON DAIRY HEIFER PERFORMANCE 18
A. Grev, S. Potts, J. Semier

EFFECTS OF WHOLE-GRAIN SUPPLEMENTATION ON HEALTH, GROWTH, AND PROFITABILITY OF PASTURE-RAISED LAME 19
A. Grev, S. Schoenian, J. Semler, D. O'Brien

DHURRIN STABILITY AND PRUSSIC ACID RELEASE IN DRIED SORGHUM PLANTS 20
S. Gruss, K. Johnson, M. Ghaste, J. Widhalm, S. Johnson, J. Holman, A. Obour, R. Aiken, M. Tuinstra

ALFALFA QUALITY IN LIMITED SUBSURFACE DRIP IRRIGATION SYSTEMS 21
K. Harmoney, F. Lamm, R. Reyes-Esteves

SEEDING WARM-SEASON ANNUAL GRASSES INTO PERENNIAL WESTERN 22
K. Harmoney, J. Guretzky

USING GPS TO IMPROVE THE ACCURACY OF FROST SEEDING 23
B. Hendrix, C. Teutsch, J. Jackson, C. Raymond

YIELD AND PROFIT MAXIMIZING ANNUAL FORAGE ROTATIONS IN THE CENTRAL GREAT PLAINS 24
J. Holman, A. Obour, Y. Assefa, L. Simon, P. Mauler

CRABGRASS RE-SEEDING AND GRAZING MANAGEMENT DEMONSTRATION.............................. 30
J. Huff

DRY FERTILIZER IMPREGNATION WITH GRAZONNEXT HL AND DURACOR HERBICIDES: BENEFITS AND LIMITATIONS........................ 32
S. Ingram

IMPACTS OF SMALL GRAIN CEREAL SILAGE MANAGEMENT ON FERMENTATION AND NUTRITIVE VALUE - A SURVEY OF NEBRASKA PRODUCERS.. 33
A. Johnson, J. Rees, B. Beckman, B. Schick, G. Lesoing, E. Laborie, K. Clark, C. Biebler, D. Redfearn, M. Drewnoski

REACHING GEORGIA FORAGE PRODUCERS IN A VIRTUAL ENVIRONMENT................................. 34
C. Knight, L. Baxter, P. Brown, R. Gates, J. Kichler, S. Morgan, L. Ray, S. Tanner, P. Sapp

IMPLEMENTING RESOURCE CONSERVATION PRACTICES ON HORSE FARMS 35
K. Lea, S. Smith

WARM-SEASON ANNUAL GRASSES AS COMPANION FORAGE CROP FOR ESTABLISHING WARM-SEASON PERENNIAL GRASSES 36
R. Lemus, J. White

FORAGE PRODUCTIVITY, WEED DENSITY, AND SOIL PROPERTIES AS AFFECTED BY GRAZING AND TILLAGE OF ANNUAL FORAGES........................ 37
P. Mauler, J. Holman, L. Simon, A. Obour, S. Johnson, K. Roozeboom

RESOURCE SELECTION BY NORTHERN BOBWHITES IN A MIXED LANDSCAPE OF EXOTIC COOL-SEASON GRASS AND NATIVE-WARM SEASON GRASS PASTURES AND FIELDS 42
D. Mitchell, P. Keyser

MIDWEST COVER CROPS COUNCIL, EMPOWERING A NETWORK OF PROFESSIONALS 43
A. Morrow

SPRING OAT AND TRITICALE FORAGE PRODUCTION AS AFFECTED BY SEEDING RATE AND NITROGEN FERTILIZER APPLICATION 44
A. Obour, J. Holman, L. Simon, P. Mauler

NITROGEN APPLICATION EFFECTS ON FORAGE SORGHUM PRODUCTION AND NITRATE CONCENTRATION 49
J. Holman, A. Obour, L. Simon, P. Mauler

UTILIZING COVER CROPS FOR FORAGE IN THE SEMI-ARID CENTRAL GREAT PLAINS 54
A. Obour, J. Holman, L. Simon, P. Mauler

OPTIMIZING TALL FESCUE HAYFIELDS THROUGH NITROGEN AND PHOSPHORUS FERTILIZATION 55
B. Pedreira, D. Sweeney, J. Farney, J. Moyer, D. Diaz

2020 RESULTS OF TIMED MECHANICAL WEED CONTROL IN EASTERN OHIO 56
T. Wiseman, C. Gelley, C. Penrose, J. McCutcheon, M. Landefeld

MOTIVATIONS AND ISSUES FACED BY LIVESTOCK PRODUCERS PRESERVING FORAGES 60
T. Picardo-Perez, R. Lemus, D. Rivera, L. Villalobos

IMPACTS OF GRAZING COVER CROPS ON ANIMAL PERFORMANCE, CROP PRODUCTION, AND SOIL HEALTH AIN EAST-CENTRAL MISSISSIPPI 61
J. Rushing, J. Lyles, B. Bass, K. Waddell

EFFECT OF SPECIES AND MATURITY ON YIELD, DIGESTIBILITY, AND PROTEIN CONTENT OF WINTER HARDY SMALL CEREAL GRAIN SILAGE 62
A. Sartin, K. Calus, M. Grabau, A. Kuhn, D. Redfearn, M. Drewnoski

NOVEL PEST MANAGEMENT STRATEGIES TO CONTROL SMUTGRASS IN SOUTHERN FORAGE SYSTEMS 63
N. Shay, L. Baxter, N. basinger, B. Schwartz

POST-WHEAT SUMMER COVER CROP EFFECTS ON CROP YIELDS AND SOIL PROPERTIES IN A NO-TILL DRYLAND CROPPING SYSTEM 64
L. Simon, A. Obour, J. Holman, M. Schipanski, S. Johnson, K. Roozeboom

EFFECTS OF THATCH DEPTH ON NATIVE WARM-SEASON GRASS (NWSG) ESTABLISHMENT 70
K. Swilling, E. Bisangwa, J. Mansfield, C. Bitler, P. Keyser

ELECTRIC FENCING FOR SERIOUS GRAZERS: A HANDS-ON PROGRAM FOR PRODUCERS AND AGRICULTURAL PROFESSIONALS 71
C. Thrailkill, C. Teutsch, M. Hayes, J. McGill

TEPARY: A DROUGHT TOLERANT FORAGE FOR THE SOUTHERN GREAT PLAINS? 72
T. Witt, B. Northup, C. Florez

Author Index

INCORPORATING PERENNIAL FORAGES INTO SOUTHERN ROW CROPPING SYSTEMS

E.D. Billman, B.T. Campbell, F.P. Reay-Jones, and G.C. Sigua[1]

Cool-season forage production in the southeastern US is dominated by annual species, requiring yearly costs for establishment. Perennials could reduce costs, but require new management practices to harmonize with row crop production. This study evaluated the efficacy of incorporating regionally-adapted cool-season perennial forages into a southeastern cotton production system. A 4×4 Latin square consisting of 1) unplanted fallow, 2) annual ryegrass ('DH-3'), 3) 50/50 red ('Southern Belle') and white clover ('Will Ladino') mixture, and 4) 33/33/33 annual ryegrass, red, and white clover mixture were established near Florence, SC in October, 2020. Plots were then strip-tilled and planted with cotton in mid-May 2021. Stand counts, weed control, and spring forage mass were assessed from February – April, while clover persistence and weedy biomass were assessed from July – October. During spring, ryegrass treatments had the greatest forage mass (>5,000 lb./acre) compared to the 50/50 clover mixture (2,000 lb./acre). During summer, treatments with clover had the lowest biomass (< 3,000 lb./acre), compared to the fallow and ryegrass treatments (> 4,000 lb./acre). Both red and white clover stands declined from spring to summer, but held steady through cotton harvest at 3 and 10 plants/ft^2, respectively. These first-year results indicated that perennial clovers can reduce weed pressure and persist over a full growing season in the Southeast without reducing cotton yield.

[1]Research Agronomist, USDA-ARS, Coastal Plains Soil, Water, and Plant Conservation Research Unit, Florence, SC (eric.billman@usda.gov); Research Geneticist, USDA-ARS, Coastal Plains Soil, Water, and Plant Conservation Research Unit, Florence, SC (todd.campbell@usda.gov); Professor of Entomology, Clemson University Pee Dee REC, Florence, SC (freayjo@clemson.edu); Research Soil Scientist, USDA-ARS, Coastal Plains Soil, Water, and Plant Conservation Research Unit, Florence, SC (gilbert.sigua@usda.gov

USING TECHNOLOGY TO MINIMIZE LABOR AND IMPROVE LIVESTOCK PRODUCTION UNDER GRAZING

C.P. Bagley, R. Palfreyman and J. Whitson[1]

Some enhanced livestock production techniques, such as deworming, external insect control, and forage digestibility modifiers, can improve animal performance, but do come at a cost of having to work animals in corrals causing added expenses and a slight reduction in animal performance. Studies have been conducted to evaluate new technologies to apply animal treatments in the field and make them more cost effective. The critical aspect is wireless connections that allow prompt animal identification that then allows the application of various treatments designed to efficiently improve animal production. A series of studies have been conducted using RFID ear tags to identify individual animals as they come to consume mineral supplements at what is called the "fly motel." The mineral feeder is solar powered, connected to the internet, and all animals tagged with individual RFID ear tags. Since cattle consume mineral supplements 2 or 3 times per week, the "fly motel" recognizes the animal through the RFID tag, and with several fitted nozzles fitted to an electric pump, can then deliver health treatments out in the pasture. Treatment for horn flies is needed about every 21 days, so if the cow has not been treated with insecticide in the last 18 days or more, the system can treat them. Tied into the National Weather Service, the system can skip spray when raining or windy. Additionally, tanks can be switched to provide dewormers twice per year as called for. Additionally, waterers can be fitted with micro-injectors to apply such treatments as biochars or ionophores based on individual animal identification.

[1]Director, Beef Research Unit, Southern University Ag Center, Baton Rouge, LA (clyde_bagley@sus.edu); Graduate Students, Southern University.

BIOCHARS IMPACTS ON PLANT GERMINATION AND GROWTH RATES WHEN LAND-APPLIED AND RUMINAL FERMENTATION RATES WHEN ADDED TO DIETS

C.P. Bagley, K.J. Han, R. Palfreyman and J. Whitson[1]

Biochars have become available with the advent of pyrolysis to generate alternative energy sources primarily from cellulosic materials. Most biochars are generated from "slow pyrolysis" generated at approximately 700 C but newer processes use temperatures of 1500 C creating a different biochar. Some literature reports benefits from the higher temperature biochar. In the first experiment, Biochar was compared to control and two levels of broiler litter (BL) using 5 legumes and 3 grasses comparing germination rates, growth rates, and root nodulation for legumes. While there were linear increases in growth rates with increases BL levels, plants did not respond to biochar as compared to controls. Plant removal rates from grazing by steers did not show treatment effects, but did indicate forage preferences for the legumes over grasses, and ryegrass and tall fescue over orchardgrass. In the second study, hays were compared for pre-bud alfalfa (23.6% CP), poor quality bermudagrass (7.0% CP) and higher quality bermudagrass (14.8%) in *in vitro* studies. Biochar suppressed digestibility of the higher quality hays, but seemed to improve digestibility low quality hay. Biochar also suppressed gas production during digestion which may indicate a reduction in the so-called "greenhouse gases" resulting from ruminal fermentation. Results with biochar in these studies were not as dramatic as those reported elsewhere. It appears there are additional factors that alter impacts of biochar and also its conversion to graphene may rule out its use as a feed additive.

[1]Director, Beef Research Unit, Southern University Ag Center, Baton Rouge, LA (clyde_bagley@sus.edu); School of Plant, Environmental, and Soil Sciences, Louisiana State University Ag Center; Graduate Students, Southern University.

USING SOCIAL MEDIA FOR FORAGE EDUCATION AND PROMOTION

Lisa Baxter and Jennifer Tucker[1]

The demands of forage livestock systems are constantly changing throughout the year and no two years are ever the same. Changing climatic conditions have exacerbated the volatility of forage production systems. For example, warm season forages and weeds are appearing further north and pest/disease outbreaks are becoming more frequent and with greater severity. It is not surprising that University of Georgia County Extension Agents ranked forage management, grazing management, weed control, and hay production as some of the greatest needs for training in the 2020 Extension Training Needs Assessment. County Extension Agents and forage-livestock producers require relevant forage information in an accessible, easy to follow format throughout the year. Social media is a powerful tool that enables forage-livestock researchers and Extension specialists to connect with a wide audience of producers, Extension personnel, colleagues, and industry partners. Online platforms provide a space for Extension personnel to deliver timely information, for researchers to provide transparency in their research program, and for industry to connect with potential collaborators. This session will provide an overview of popular social media platforms (Facebook, Twitter, Instagram, TikTok, and YouTube) and how to transform impactful research into effective social content.

[1]Assistant Professor and Extension Forage Specialist, Crop and Soil Science Dept., University of Georgia, Tifton, GA; (baxterl@uga.edu); Associate Professor, Animal and Dairy Science Dept., University of Georgia, Tifton, GA; (jjtucker@uga.edu).

IMPACT OF MANAGEMENT ON BERMUDAGRASS STEM MAGGOT DAMAGE

Lisa Baxter, Bill Anderson, and Will Hudson[1]

Since its discovery in South Georgia in the summer, 2010, bermudagrass stem maggot (BSM; *Atherigona reversura* Villenueve) has severely damaged bermudagrass (*Cynodon dactylon*) throughout the Southeastern US. Although the degree of damage depends on the bermudagrass variety, latitude of the farm, and time of year, producers have reported up to an 80% yield loss in late summer. Although strategic pesticide use can effectively suppress BSM populations, it is critical that work continue to address the lack of diversity in available BSM management tools. Six bermudagrass varieties (Alicia, Coastal, Coastcross-II, Russell, Tifton 44, and Tifton 85) were evaluated in a split-block experiment at the Darrell William Research Farm in Moultrie. GA. Initially the experiment was designed to determine the impact of the BSM on the most popular varieties grown in the Southeast US. However, we observed that several small management changes led to big improvements in forage production and consequently reduced the impact of BSM. Hay production increased substantially, from 5,000 lb./ac to 11,000 lb./ac in just three years ($P < 0.01$). This productivity was the result of management changes such as cutting on time, adjusting the harvest height, and better distributing fertilizer across the season. During this period, a reduction in BSM damage was also observed ($P < 0.01$). Good forage management plays a larger role in BSM than previously thought. Future research will identify which specific forage management strategy enables the forage to best resist BSM damage. Funding for this work was provided by the Georgia Beef Commission.

[1]Assistant Professor and Extension Forage Specialist, Crop and Soil Science Dept., University of Georgia, Tifton, GA; (baxterl@uga.edu); Research Geneticist, USDA-ARS, Tifton, GA (bill.anderson@ars.usda.gov); Extension Specialist, Entomology Dept., University of Georgia, Athens, GA (wghudson@uga.edu).

PATCH-BURN AND ROTATIONAL GRAZING EFFECTS ON PERFORMANCE OF BEEF HEIFERS AND PASTURE PRODUCTIVITY

M. Berry, P. Keyser, R. Smith and E. Bisangwa[1]

Patch-burn grazing (PBG) uses prescribed fire to distribute cattle grazing pressure within a pasture and can be an efficient management strategy to replace more labor-intensive approaches, such as rotational grazing (ROT). Still, PBG has not been explored in eastern forage systems. Therefore, we implemented an experiment at three sites, two in Kentucky and one in Tennessee, using either weaned or yearling heifers. These animals were randomly assigned to 25-ac pastures managed with either PBG or ROT with each divided into three patches/paddocks. Pastures had been seeded with a native warm-season grass mix [big bluestem (*Andropogon* gerardii) and indiangrass (*Sorghastrum nutans*)] three years prior to the initiation of grazing. Cattle were weighed yearly before initial grazing and again after final removal. Forage samples were taken within actively grazed ROT paddocks (n = 16) and within burned (n = 8) and unburned (n = 8) PBG grazing paddocks. Tiller counts were conducted yearly during mid-summer. The average daily gain was comparable in treatments at DREC ($P < 0.86$) and QUICK ($P < 0.66$) but interacted by treatment and year ($P < 0.01$) at BGAD. Inconsistently, grazing days differed by treatment at both DREC ($P <0 .05$) and QUICK ($P <0 .001$) and declined ($P < 0.001$) in all three years at BGAD. Although forage mass declined through the summer at all locations and tiller counts declined at one site over the course of the study, both forage mass ($P < 0.77$) and tiller counts ($P < 0.75$) were similar between treatments.

[1]PhD candidate, University of Tennessee (mberry@vols.utk.edu), Center for Native Grasslands Management, University of Tennessee (pkeyser@utk.edu), University of Kentucky (raysmith1@uky.edu), Center for Native Grasslands Management, University of Tennessee (ebisangw@vols.utk.edu).

EXTRACTION OF LIGNIN MONOMERS FOR UPLC ANALYSIS

M.L. Boss, J.R. Strickland, M.J. Aguerre[1]

Lignin is a structural component that is bound to cellulose and hemicellulose in the plant's cell wall. While it allows the plant to retain rigidity, it is also indigestible to animals. The phenolic polymer is made up of 3 primary monomers: sinapyl alcohol, p-coumaryl alcohol, and coniferyl alcohol. These monomers further polymerize into p-hydroxyphenyl (H), guaiacyl (G), and syringal (S). These polymerized subunits and other precursors may be characterized utilizing Ultra Performance Liquid Chromatography (UPLC). This study was conducted to characterize the lignin components that may be downregulated in reduced lignin alfalfa varieties and to propose UPLC as a method to qualify lignin content. Samples to be tested were grown between March 2021 and October 2021, with five harvests total. Samples were a part of a 20-plot complete randomized block design with either (1) reduced lignin alfalfa monoculture, (2) novel endophyte-free tall fescue monoculture, (3) conventional alfalfa monoculture, (4) reduced lignin/tall fescue mixture, or (5) conventional alfalfa/tall fescue mixture. We hypothesized the reduced lignin variety would exhibit significantly lower levels of total lignin, and the S component of lignin would be downregulated. Components were characterized and quantified utilizing UPLC with both sample standards and research grown samples. UPLC data will be analyzed and outlined to illustrate lignin quality and components in both sample types.

[1]Graduate student, Animal and Veterinary Science Department, Clemson University, Clemson, SC (mbrizen@g.clemson.edu); Nutritional Toxicologist, Animal and Veterinary Sciences Department Clemson University, Clemson, SC (jrstric@clemson.edu); Ruminant Nutritionist, Animal and Veterinary Sciences Department, Clemson University, Clemson, SC (maguerr@clemson.edu).

GROWTH AND ENSILING OF REDUCED LIGNIN ALFALFA IN THE SOUTHEASTERN US

M.L. Boss, M.J. Aguerre, J.R. Strickland[1]

Reduced lignin varieties are a relatively novel advancement in alfalfa genetics. A reduction in lignin suggests higher digestibility in livestock, increased harvest intervals, while maintaining or increasing forage yield. This study was conducted to determine if quality factors, establishment rates, yield, and silage quality were affected by the reduction of lignin in alfalfa when grown in the Southeastern United States. Samples to be tested were grown between March 2021 and October 2021, with five harvests total. Samples were a part of a 20-plot complete randomized block design with either (1) reduced lignin alfalfa monoculture, (2) novel endophyte-free tall fescue monoculture, (3) conventional alfalfa monoculture, (4) reduced lignin/tall fescue mixture, or (5) conventional alfalfa/tall fescue mixture. We hypothesized that reduced alfalfa plot yields would match or exceed those of the conventional alfalfa yields in both monoculture and tall fescue mixture. In addition, collected samples were ensiled at either low or high dry matter content and with or without lactic acid bacteria inoculant. This was done in order to highlight any ensiling differences between reduced lignin monoculture or tall fescues mixtures against conventional alfalfa. Additional harvest cycles may be needed to accurately quantify establishment rates through time.

[1]Graduate student, Animal and Veterinary Science Department, Clemson University, Clemson, SC (mbrizen@g.clemson.edu); Ruminant Nutritionist, Animal and Veterinary Sciences Department, Clemson University, Clemson, SC (maguerr@clemson.edu); Nutritional Toxicologist, Animal and Veterinary Sciences Department Clemson University, Clemson, SC (jrstric@clemson.edu).

MEASURING IMPACT OF FERTILIZATION ON BROOMSEDGE ENCROACHMENT IN HAYFIELDS USING REMOTE SENSING

J.A. Bush, C.D. Teutsch, J.C. Henning, and S.R. Smith[1]

This study investigates using remote sensing to quantify warm- and cool-season grasses in a hay field. Alternant dormancy periods create a color dichotomy between cool and warm season grasses throughout the year. This may allow us to use green canopy cover (GCC) software to measure percentages of cool and warm season grasses. This study was conducted in a hay field that was originally dominated by cool-season grasses such as tall fescue (*Schedonorus arundinaceus*). Over time, this field was invaded by broomsedge (*Andropogon virginicus*), a native warm-season grass that is known to be an indicator of poor soil fertility. Soil tests for this field revealed a severe potash deficiency. Thirteen different fertilizer regimens were applied with four replications each in a randomized complete block design. Images were captured with a Nikon D750 DSLR camera and a DJI Phantom 4 Pro V2.0 drone throughout 2020 and analyzed for GCC percentage using Canopeo for MATLAB. Broomsedge-dominated plots had strong green color (high GCC) during the warmest part of the year and a faded green or brown color (low GCC) during the cooler months. High NPK fertilizer treatments showed the highest GCC values throughout the year, including cooler months when compared to low fertilization plots. This relationship shows a reduction of warm season grasses via fertilization. This method of collecting data was both time and labor efficient and yielded adequate and useful data.

[1]Graduate Student – Jordyn Bush, Univ. of KY (jordynbush@uky.edu); S. Ray Smith, Extension Professor, Univ. of KY; Jimmy Henning, Extension Professor, Univ. of KY; Chris Teutsch, Extension Professor, Univ. of KY – Princeton (UKREC).

EVALUATION OF REMOTE SENSING TECHNIQUES TO MEASURE CRABGRASS ENCROACHMENT IN TALL FESCUE STANDS

J.A. Bush, C.D. Teutsch, J.C. Henning, and S.R. Smith[1]

Climate change is altering farm management, and forage production is no exception. The ratio of warm- to cool-season (C3) grasses may change with warming temperatures across the US, especially in transition zones like Kentucky. Alternate growth and dormancy periods create a color dichotomy between cool and warm season grasses which can be detected in photographs from drones and digital cameras. The objective of this study was to use remote sensing to measure ratios of tall fescue and crabgrass throughout the year. Tall fescue mosaics of 1m x 1m squares were 0, 25, 50, 75, or 100% converted to crabgrass, creating a known ratio of cool and warm season grasses. These mosaic plots were photographed using a Nikon digital camera or a DJI Phantom V2.0 drone in spring, summer, fall, and winter. Photos were analyzed for green canopy cover using Canopeo for MATLAB. The highest correlation between expected and measured values was seen in the winter sampling, with regression equations for the drone images and digital camera images being y=0.49x + 0.24 (R^2=0.95) and y=0.56x+0.97 (R^2=0.95) respectively. Future applications of remote sensing in pasture analysis include measuring and monitoring cool and warm season grass ratios to observe how they may change over time, and to help farmers make management decisions based on these measured changes.

[1]Graduate Student – Jordyn Bush, Univ. of KY (jordynbush@uky.edu); S. Ray Smith, Extension Professor, Univ. of KY; Jimmy Henning, Extension Professor, Univ. of KY; Chris Teutsch, Extension Professor, Univ. of KY – Princeton (UKREC).

MEASURING PROPORTION OF LIVE AND DEAD VEGETATION IN PASTURES USING REMOTE SENSING

J.A. Bush, C.D. Teutsch, J.C. Henning, and S.R. Smith[1]

Anecdotal evidence from Kentucky farmers indicates an increase in warm season species in previously cool-season dominated pastures. In this study, we aim to evaluate digital photography as a method for monitoring these changes in pasture composition. Due to alternate periods of seasonal dormancy, cool and warm season species display differences in color throughout the year. During cooler periods cool season grasses retain their green color while warm season species lose this coloring. These color differences are readily visible in photographs. This visibility has inspired this study. This study analyzed the use of digital camera imagery and drone imagery to measure strips of brown (dead) and green (alive) grass in a pasture. A randomized complete block design with three replications was used. Treatments were 25, 50, and 75% of total area sprayed with non-selective herbicide. In April of 2020, strips were sprayed out of paddocks that had been planted to wheat. Ground measurements were used to confirm treatment percentages. Thirty digital images were collected in a random pattern in each paddock and imagery was repeated twice per paddock. Green canopy cover (GCC) was measured using Canopeo for MATLAB. A regression equation for all data collected graphed against the expected GCC values has an r^2=0.87 (y=0.66x+0.27). This initial study demonstrates that digital imagery may be a time-efficient, cost-effective method for monitoring the long-term shifts in the ratios of cool- to warm-season grasses in pastures in the transition zone of the United States.

[1]Graduate Student – Jordyn Bush, Univ. of KY (jordynbush@uky.edu); S. Ray Smith, Extension Professor, Univ. of KY; Jimmy Henning, Extension Professor, Univ. of KY; Chris Teutsch, Extension Professor, Univ. of KY – Princeton (UKREC).

FORAGE PRODUCTION AND NUTRITIONAL QUALITY OF TALL FESCUE AND ALFALFA UNDER LONG TERM WARMING AND SEVERE DROUGHT

K.L.M. Catunda, A.C. Churchill, S.A. Power, and B.D. Moore[1]

Changes in climate, such as increases in temperature and severe drought events, have been identified as one of the main concerns for temperate pasture-based systems in many regions of the world. Understanding the impacts of future climates on forage systems can inform adaptation strategies that help mitigate potential negative effects. This study investigated the impacts of future climate change scenarios of warming and severe drought, as independent factors and in combination, on forage production and nutritional quality of two widely used temperate pasture species, tall fescue (*Festuca arundinacea*) and alfalfa (*Medicago sativa*). We exposed these species to two-years of continuous warming (+3 °C above ambient temperature) and two winter/spring periods of severe drought (60% rainfall reduction) at the Pastures and Climate Extremes (PACE) experimental field facility in southeast Australia. Warming and drought significantly reduced productivity of both species and slightly increased nutritional quality in tall fescue, while leaving alfalfa quality largely unchanged. The combination of warming and severe drought resulted in effects that varied from additive to less than additive. The results of our study indicate that even though forage quality was not reduced, observed large decreases in pasture production will present a challenge for maintaining livestock productivity rates under future climate conditions.

[1]PhD student, Hawkesbury Institute for the Environment, Western Sydney University (zootecnistakaren@hotmail.com); Postdoctoral Fellow, Department of Ecology, Evolution and Behavior, University of Minnesota (achurchi@umn.edu); Professor, Hawkesbury Institute for the Environment, Western Sydney University (s.power@westernsydney.edu.au); Senior Lecturer, Hawkesbury Institute for the Environment, Western Sydney University (b.moore@westernsydney.edu.au).

EXTREME SUMMER DROUGHT IMPACTS ON OLD-FIELD FORAGE PRODUCTION AND NUTRITIONAL QUALITY

A.C. Churchill and F.I. Isbell[1]

A future that includes long periods of summer drought will require creative solutions and considerations for alternative management strategies in limiting productivity losses in grasslands, including the plant species comprising forage. To examine underlying factors that dictate grassland responses to drought, we implemented a three-month extreme drought via precipitation reduction across a range of old-fields abandoned between 6 and 94 years from agricultural use. This old-field age chronosequence includes grasslands with contrasting plant species diversity and soil fertility, among other characteristics, at the Cedar Creek Ecosystem Science Reserve in MN. Our 43% precipitation reduction treatment resulted in soil water content that was 18% less than ambient conditions. The impact of this applied drought ranged from a -55% (50 years abandoned) to +30 % (94 years) change in aboveground biomass among the different old-fields, with drought reducing biomass by 20% overall. Soil samples from the same fields show accumulations of soil carbon with time since abandonment, as well as changes in soil compaction, which alongside variability in plant species diversity, may contribute to the range of drought response observed among old-fields. These measurements and analyses evaluate the underlying mechanisms that land managers may consider in predicting and reducing losses in productivity associated with drought conditions among diverse types of grassland communities.

[1] Postdoctoral Fellow, Department of Ecology, Evolution and Behavior, University of Minnesota (achurchi@umn.edu); Associate Professor, Department of Ecology, Evolution and Behavior & Associate Director, Cedar Creek Ecosystem Science Reserve, University of Minnesota (isbell@umn.edu).

IMPACT OF BMR TRAIT ON DECOMPOSITION RATE OF SORGHUM-SUDANGRASS

B.P. Dooley[1], J.A. Bush[2], C.D. Teutsch[3], and B.L. Hendrix[4]

Sorghum-sudangrass (*Sorghum bicolor* var. bicolor x bicolor var. sudanense) can provide high quality summer grazing. Some varieties possess the brown midrib (BMR) trait which results reduced lignin. This trait has been linked to higher digestibility and animal performance. If microbes in the rumen can digest BMR sorghum-sudangrass more completely, then soil macro/micro flora and fauna may do so as well. This could result in nutrients being returned to the soil faster from plants containing the BMR trait. The objective of this study was to determine the decomposition rate of BMR and non-BMR sorghum-sudangrass. Sorghum-sudangrass with and without the BMR trait was placed in litter decomposition bags as whole plants or divided into leaves and stems. A composite sample was taken when bags were loaded to determine initial dry matter. Loaded bags were then placed on the soil surface in a pasture and collected at 1, 2, 3, 4, 6, 8, 10, 12, 14, or 16 weeks after placement. Upon collection plant material was dried for 3-days at 55C in a forced air oven. Dry weights at each collection date were subtracted from the initial dry weight to determine total DM loss. The BMR trait did not impact dry matter loss in the leaves. Stems possessing the BMR trait lost dry matter at a greater rate resulting in dry matter losses at 14 weeks of 78.5 and 68.6% and 59.1 and 47.6% for the BMR and non-BMR varieties in trials 1 and 2, respectively. Whole plants showed limited differences in dry matter loss at 14 weeks after placement.

[1]USDA-NIFA Intern, UK Research and Education Center, Princeton, KY (brandon.dooley@uky.edu); [2]Graduate Research Assistant, UK Plant and Soil Science Dept., Lexington, KY; and [3]Extension Associate Professor and [4]Senior Research Technician, UK Research and Education Center, Princeton, KY.

INVASIVE SPECIES ON KANSAS GRASSLANDS

W.H. Fick[1]

Grasslands in Kansas consist primarily of native rangeland and tame pasture. These grasslands total over 18 million acres and contribute substantially to the livestock and forage industry. Invasive plant species typically reduce forage production and upset the ecological integrity of the grasslands. Three key invasive species in Kansas grasslands include sericea lespedeza (*Lespedeza cuneata*), Old World Bluestems (*Bothriochloa* spp.), and saltcedar (*Tamarix ramosissima*). Sericea lespedeza infests over 600,000 acres, primarily in the eastern two-thirds of the state, making it a Category C noxious weed in the state. Herbicides containing triclopyr or metsufuron are the most effective at reducing sericea lespedeza populations, but a large seedbank of viable seed allows these populations to recover in 2-3 years. Grazing by sheep and goats can help reduce stands of sericea lespedeza and repeated mowing can eliminate seed production and reduce vigor. Fire enhances seed germination, but repeated late summer burning has nearly eliminated seed production and may reduce stands. The Old World Bluestems (OWB's) consist of Caucasian bluestem (*B. bladhii*) and yellow bluestem (*B. ischaemum*). These species exist in every county in Kansas, typically spreading from roadside seedings or plantings made during the Soil Bank days in the late 1950s and early 1960s. The OWB's are less palatable than most native grasses and are invading grazed and hayed grasslands. Glyphosate and imazapyr are used to control OWB. Saltcedar occurs primarily along the major river systems in Kansas. Cut-stump, basal bark, and foliar herbicide treatments help control saltcedar. Integrated systems generally enhance the control of invasive species.

[1]Extension Rangeland Management Specialist, Department of Agronomy, Kansas State University, Manhattan, KS (whfick@ksu.edu).

CHLOROPHYLL FLUORESCENCE AS AN INDICATOR OF PLANT FORAGE STRESS

Sarah C. Goslee[1]

Chlorophyll fluorescence (CF) is an effective measure of plant stress in other systems, and is fast and non-destructive, requiring only a few minutes per sample and conducted completely in the field. As a measure of plant stress in forages, it might provide both an indication of forage production and even quality and a rapid protocol for quantifying species response to growing conditions in different field trials. This technique has rarely been used in pastures, so a pilot study was developed to investigate the potential use of CF in grazing systems. The study was conducted on a forage species mixture trial at Penn State's Rock Springs Experimental Farm, University Park, PA, USA. Orchardgrass, white clover, and chicory were planted in autumn 2019, alone and in combination. Chlorophyll fluorescence was measured with a hand-held instrument (Pocket Pea, Hansatech Instruments Ltd, UK) on five dark-adapted leaves of each planted species per plot. Measurements were taken four times a year for two years at eight-week intervals. A mixed-effects analysis of chlorophyll fluorescence identified a strong species effect, a weaker treatment effect, and no effect of sampling date. However, 2021 was much wetter than 2020 during the growing season, resulting in strong inter-year patterns. The differences across species, across treatments, and across years demonstrate that chlorophyll fluorescence has the potential to be a useful indicator of stress in forage plants, and that further investigation is justified.

[1]Research Ecologist, USDA-ARS Pasture Systems and Watershed Management Research Unit, University Park, PA (sarah.goslee@usda.gov).

COMPAIRISON OF FIVE METHODS OF BOTANICAL ANALYSIS ACROSS KENTUCKY

E.E Gotsick, S.R. Smith, J.C. Henning, and C.D. Teutsch[1]

Botanical composition shows ecological diversity, provides an indication of forage quality and palatability, and identifies the species with potential toxicity like tall fescue. The objective of this study is to determine the most accurate and efficient method for measuring botanical composition in pastures. This study is being conducted on six cattle farms across Kentucky, two pastures per farm, and six sampling dates over two years representing four of the five geographical regions of Kentucky: Bluegrass Region, The Cumberland Plateau, the Western Coal Field, and the Pennyroyal Region. The different methods of botanical composition used in this study are: Step Point, Visual Estimation, Occupancy, Point Quadrat, and Pasture Scoring. Three of the methods (visual estimation, occupancy, and point quadrat) will involve 20 sample locations randomly selected across each pasture and will be repeated from digital images taken at each location. The step point method involves 100 point measurements in a random grid pattern and the pasture score method is a modification of the USDA-NRCS Pasture Health Score Card providing general categorical rating of each pasture. Measurement dates include Spring (April/ May) to establish a baseline of the cool season grass growth. Summer (July/August) to determine the pasture composition change due to warm season grasses/ forbs. Fall (September/ November) contrast dying warm season grasses and cool season grasses. The first year of data has been collected and analysis is ongoing. The overall goal of this study is to determine an accurate and efficient method of botanical composition for researchers, practitioners, and students.

Graduate Student- Echo Gotsick, Univ. of KY (echo.gotsick@uky.edu), S. Ray. Smith, Extension Professor, Univ. of KY, Jimmy Henning, Extension Professor, Univ. of KY, Chris Teutsch, Extension Professor, Univ. of KY – Princeton (UKREC).

EFFECT OF AN IMPROVED GRAZING MANAGEMENT SYSTEM ON DAIRY HEIFER PERFORMANCE

A.M. Grev[1], S.B. Potts[1], J.W. Semler[1]

The objectives of this study are to investigate the effects of improved grazing management on pregnant dairy heifer performance, as well as its economic feasibility relative to a conventional management system. Heifers at the Central Maryland Research & Education Center dairy were utilized for this study on a rolling basis, with heifers added following pregnancy confirmation and removed 3 weeks prior to calving. Heifers were divided into two treatment groups which varied in size throughout the study period (15 to 22 per group) but were kept consistent between treatments. Heifers in the CON group received a daily TMR and had access to one continuously managed 5-acre perennial pasture. Heifers in the ROT group were rotationally grazed across 19 acres of both perennial and annual pastures subdivided into approximately 0.6-acre paddocks (1-2 d rotation); heifers also received a daily corn/mineral mix (1 lb/head/d) to ensure dietary requirements were being met. Both groups of heifers were measured on a bi-weekly basis throughout the grazing season to determine BW, BCS, and hip height. Results are presented for Apr 7 through Nov 24, 2021. Initial BW, BCS, and hip height did not differ between treatment groups. Final BW and hip height did not differ, but final BCS was greater for CON heifers (3.7) compared to ROT heifers (3.5; $P \leq 0.05$). For both groups, heifer average daily gain (ADG) varied considerably among weigh periods, but ADG was greater for CON compared to ROT ($P \leq 0.05$). These results indicate a slight growth disadvantage for the grazing system; however, heifer performance was still good and expenses for this system may be reduced compared to a conventionally managed system.

[1]University of Maryland Extension, Keedysville, MD.

EFFECTS OF WHOLE-GRAIN SUPPLEMENTATION ON HEALTH, GROWTH, AND PROFITABILITY OF PASTURE-RAISED LAMB

A.M. Grev[1], S.G. Schoenian[1], J.W. Semler[1], D.J. O'Brien[2]

The objectives of this project were to compare the health, growth, and market value of lambs raised on pasture with and without whole-grain supplementation to determine if the additional energy provided through supplementation will improve the profitability of pasture-based lamb production. On June 15, Katahdin ram lambs (n=79 in 2020; n=85 in 2021) arrived at the Western Maryland Research & Education Center. After a 10-d acclimation period, lambs were allocated into two equal treatment groups based on age, weight, birth type, and fecal egg count (FEC). Lambs in the PAST group rotationally grazed 5 acres of high quality, mixed pasture for 102 d. Lambs in the SUPPL group grazed similar pastures and were hand-fed a daily energy supplement at the rate of 1 lb of whole barley per head. Both groups were handled bi-weekly to determine BW, BCS, and FAMACHA© scores. Individual fecal samples were collected upon arrival and at 1-2 additional time points each year. Data was analyzed using the MIXED procedure of SAS, with statistical significance set at $P \leq 0.05$. Average daily gain (ADG) varied considerably among weigh periods and standard deviations were typically large, indicating wide variation in individual performance. Across both years, final BW did not differ but the SUPPL lambs had higher ADG (0.32 lb/d) and final BCS (2.8) compared to the PAST lambs (0.27 lb/d and BCS 2.6; $P < 0.01$). Final FEC and FAMACHA© scores were lower for the PAST lambs ($P < 0.05$) in 2020 but similar between groups in 2021. These results indicate a slight growth advantage for the SUPPL lambs; however, the value of the additional gain likely would not cover feed expenses.

[1]Univ. of Maryland Extension, Keedysville, MD; [2]Virginia State University, Petersburg, VA

DHURRIN STABILITY AND PRUSSIC ACID RELEASE IN DRIED SORGHUM PLANTS

Shelby M. Gruss[1], Keith D. Johnson[1], Manoj Ghaste[2,3], Joshua R. Widhalm[2,3], Sandy K. Johnson[4], Johnathon D. Holman[5], Augustine Obour[6], Robert M. Aiken[4], Mitchell R. Tuinstra[1]

Sorghum (*sorghum bicolor* (L) Moench) is an important forage crop throughout the United States. Currently all sorghum on the market contains the cyanogenic glucoside dhurrin. When dhurrin breaks down, it is released as hydrogen cyanide (HCN) gas, which is extremely toxic to livestock. Utilizing sorghum as hay has been reported to decrease dhurrin content by 50 percent but this has not been clearly substantiated. The objective of this research was to evaluate dhurrin content and HCN release of dried sorghum tissue samples. Dried plant tissues from conventional and dhurrin-free sorghum lines were evaluated for dhurrin content and HCN release over time. Dhurrin content remained stable in conventional sorghum tissue samples for at least two months; however, HCN production declined over the two months when dried sorghum tissues were rehydrated with water. Rehydration of dried sorghum tissues with rumen fluid led to rapid release of HCN. The dhurrin-free sorghum hybrid did not produce detectable levels of dhurrin or release HCN at any time points or rehydration methods. The drying process during haymaking does not decrease dhurrin content or reduce the risk of rapid HCN release when rehydrated with rumen fluid.

[1]Department of Agronomy, Purdue University, West Lafayette, IN 47907, USA, [2]Department of Horticulture and Landscape Architecture, Purdue University, West Lafayette, IN 47907, USA, [3]Center for Plant Biology, Purdue University, West Lafayette, IN 47907, USA, [4]Northwest Research-Extension Center, Kansas State University, Colby, KS 67701, USA, [5]Southwest Research-Extension Center, Kansas State University, Garden City, KS 67846, USA, [6]Agricultural Research Center, Kansas State University, Hays, KS 67601, USA.

ALFALFA QUALITY IN LIMITED SUBSURFACE DRIP IRRIGATION SYSTEMS.

Keith Harmoney, Freddie Lamm, and Rocio Reyes-Esteves[1]

Alfalfa (*Medicago sativa* L.) has historically been an important crop in western Kansas to supply the feedlot and dairy industries with a high quality forage protein source. Depending on the year, western alfalfa production uses approximately 5-10% of Kansas' irrigated acres. Subsurface drip irrigation (SDI) systems under alfalfa are viewed as an alternative to center pivot systems in this region to increase efficiency of water use and to mitigate concerns with declining groundwater supplies. Prior SDI research recommended that alfalfa be maintained at an irrigation level equal to 85% of evapotranspiration (ET) levels to maintain available soil moisture and deter production declines during severe drought periods, although forage yield and quality were maintained outside of these drought periods at 70% ET levels. Alfalfa quality increased as soil conditions dried at lower irrigation levels and at greater distances away from the dripline. Follow-up studies that supplied water to alfalfa at 85% of ET before the first and last harvests, but limited irrigation to 25% to 70% of ET before harvests in between, showed conflicting quality trends compared to the prior study. Irrigation level had little effect on forage quality. Forage crude protein and neutral detergent fiber digestibility was greater directly over driplines, while in situ digestibility was greater for forage away from the dripline. Fiber components showed no consistent trends due to distance from the drip line or the irrigation treatment. The main common trends with all experiments were that forage yields significantly declined and forage quality generally increased with later harvests.

[1]Keith Harmoney, Kansas St. Univ. Agricultural Research Center, Hays, KS, harmone@ksu.edu; Freddie Lamm and Rocio Reyes-Esteves, Kansas St. Univ. Research and Extension Center, Colby, KS, flamm@ksu.edu; rocior@ksu.edu.

SEEDING WARM-SEASON ANNUAL GRASSES INTO PERENNIAL WESTERN WHEATGRASS PASTURE

Keith Harmoney and John Guretzky[1]

A reduction in total acreage of pastureland from conversion to cropland has resulted in a decline of beef cow numbers in the Great Plains. One method to mitigate the decline in cow numbers is to increase carrying capacity of remaining pastureland acres. A study was conducted to introduce warm-season annual grasses into perennial cool-season grass pastures to increase dry matter production during mid-summer when perennial cool-season grasses would be most dormant. Following spring wheatgrass harvest, three warm-season annual grasses (sorghum-sudangrass, crabgrass, and teff) were no-till drilled at three seeding rates into western wheatgrass pasture and then fertilized. Warm-season grasses were harvested 90 days after planting. Western wheatgrass yields prior to seeding warm-season annual grasses were 2230 and 1600 lbs/acre in 2020 and 2021, and were not different among treatments. Following wheatgrass harvest in June, precipitation adequate for germination of annual warm-season grasses was lacking. Sorghum-sudangrass was the only species with adequate stands following a significant precipitation event in both years. At 90 days after seeding, sorghum-sudangrass plots averaged 6600 and 810 lbs/acre additional forage in 2020 and 2021. Productive crabgrass and teff plants were rare. In spring of 2021, plots with sorghum-sudangrass in 2020 had western wheatgrass yields that averaged 400 lb/acre less than unseeded plots and the averages of the failed teff and crabgrass seedings. Seeding sorghum-sudangrass into western wheatgrass did slightly reduce wheatgrass yield the following year, but significantly increased total forage production.

[1]Keith Harmoney, Kansas St. Univ. Agricultural Research Center, Hays, KS, harmone@ksu.edu; John Guretzky, Univ. of Nebraska Dept. of Agronomy and Horticulture, Lincoln, NE, jguretzky2@unl.edu.

USING GPS TO IMPROVE THE ACCURACY OF FROST SEEDING

B.L. Hendrix, C.D. Teutsch, J. Jackson, and C.P. Raymond[1]

Guidance utilizing GPS has long been used for various operations in row crop agriculture. However, the high cost of these systems has limited their use in low-input forage and livestock operations. Reduced prices and the availability of used guidance systems have the potential to increase the use of precision agriculture in pastoral settings. In the past, frost seeding often resulted in areas that received no seed and areas that were double seeded. The objective of this experiment was to evaluate the impact of using a guidance system on the uniformity of seed dispersal. This study was conducted at the University of Kentucky's Research and Education Center, located in Princeton, KY in 2019 and 2021. The experimental design was a randomized complete block with four replications. Four pastures ranging from 6.2 to 10.6 acres were mock seeded using a UTV equipped with a Raven Cruizer II guidance system (Raven Industries, Sioux Falls, SD). The guidance system was initiated, but covered with a bag, and the four pastures were driven by sight alone. This process was then repeated utilizing the guidance system. Frost seeding without GPS guidance resulted in a 49% and 21% overlap in 2019 and 2021, respectively. At an overseeding cost of $36/acre and an average overlap of 35%, the cost of a guidance system could be recouped in as little as 119 acres. The results of this study indicate that GPS guidance systems have the potential to improve the uniformity of seed dispersal, thus reducing the cost of frost seeding.

[1]Senior Research Technician, University of Kentucky Research and Education Center, Princeton, KY (brittanyhendrix@uky.edu); Extension Associate Professor, University of Kentucky Research and Education Center, Princeton, KY (chris.teutsch@uky.edu); Assistant Extension Professor, University of Kentucky Department of Biosystems and Agricultural Engineering, Lexington, KY (joshjackson@uky.edu); Grain Crops Extension Associate, University of Kentucky Research and Education Center, Princeton, KY (conner.raymond@uky.edu).

YIELD AND PROFIT MAXIMIZING ANNUAL FORAGE ROTATIONS IN THE CENTRAL GREAT PLAINS

J.D. Holman[1], A.K. Obour[2], Y. Assefa[1], Logan Simon[2], and P.S. Mauler[1]

Abstract

Forage-based rotations with greater forage nutritive values and efficient resource utilization are needed to increase forage availability in the central Great Plains. The objectives of the current study were to quantify forage yield, nutritive value and net return from individual crops and overall crop rotation sequence of winter triticale [×*Triticosecale Wittm.* ex A. Camus (Secale ×Triticum)], forage sorghum [*Sorghum bicolor* (L.) Moench], and spring oat (*Avena sativa* L.). The study was conducted near Garden City, KS from 2013 - 2020. There were six treatments including; continuous sorghum (S-S) no-till (NT), triticale sorghum-sorghum-oat (T/S-S-O) reduced till (RT) and NT, triticale/sorghum-sorghum-sorghum-oat (T/S-S-S-O) RT and NT, and triticale-sorghum-oat (T-S-O) NT. Results showed crude protein concentration was 23-31% greater for T-S-O and T/S-S-O compared with S-S. The S-S treatment had greater fiber content and least digestibility, Ca, and P concentrations compared with the remaining forage rotations. Annualized cost of production was 9-32% greater for forage sorghum dominated NT systems [T/S-S-S-O, S-S (NT)] compared with T-S-O and T/S-S-O (RT) rotation. T/S-S-S-O (RT) was among the top in net return and nutritive value due to greater forage sorghum biomass production. We concluded forage producers should consider productivity and market value, in addition to nutritive value and cost of production in selecting crops for a profitable forage-based rotation system.

Introduction: Integrating annual forages into traditional grain crop rotations increased productivity, weed control, water use efficiency, and soil health. A forage-based cropping system can lower cost of production and provide more stable net return than continuous grain cropping. Even though forage rotation systems are advantageous compared with grain-only based systems; forage productivity, nutritive value, and cost of production varies by forage species and information is limited for which forage crop and crop rotation sequence optimizes nutritive value and net return in the central Great Plains.

Forage sorghum, oat, and triticale are among the most adapted annual forages in the central Great Plains. Forage sorghum is an annual warm-season crop, drought- and heat-tolerant, and adapted to semi-arid regions of the world. Triticale is an annual cool season crop with winter hardiness and high forage nutritive value, and it is a better alternative to wheat grown for forage in the Great Plains (Holman et al., 2010, 2021). Like triticale, oat is a cool-season crop grown for forage and it is highly nutritious as livestock feed. A rotation of warm-season and cool-season forage crops contributes to improved use of precipitation and resources available across different seasons in the year compared with a continuous cropping of either of these crops (Holman et al., 2020). Furthermore, growing forages during different periods of the year in semi-arid

---

[1]Southwest Research-Extension Center, Kansas State University, Garden City, KS; Agricultural Research Center-Hays, Kansas State University, Hays, KS.

environments with variable precipitation can help reduce forage yield variability and ensure availability of sufficient forage resources to meet livestock demand. Forage nutrient value and biomass production from cool-season and warm-season forages are different. For example, warm-season grasses tend to have high biomass productivity and cool season grasses have better forage nutrient value, and therefore they might become complementary when grown in rotation.

For a forage rotation system to be sustainable, it has to be productive, efficient in resource use, provide nutritious forage and, overall, it has to be profitable (Holman et al., 2020). Often, a productive and resource use efficient rotation system may be profitable at the enterprise level, but if the forage nutritive value is low, purchased (often expensive) supplemental feed may be needed to meet animal production goals. The objectives of the current study were to quantify forage yield, nutritive value and net return from individual crops (winter triticale, forage sorghum, and spring oat) and crop rotation sequences of that same study.

Material and Methods: Forage yield, nutrient analysis and economic data presented herein were from a study conducted at the Southwest Research-Extension Center near Garden City, KS from 2013 through 2020. The experimental design was a randomized complete block. Crops in the rotation were winter triticale (T), forage sorghum (S), and spring oat (O). An incomplete factorial combination of four rotations of these crops (S-S, T/S-S-O, T/S-S-S-O, and T-S-O) and two tillage [(no-tillage (NT), reduced tillage (RT)] practices formed the six treatments of the study. These were (1) continuous forage sorghum rotation with no-tillage (S-S NT); (2) winter triticale/double crop forage sorghum-forage sorghum-spring oat rotation with no-tillage (T/S-S-O NT); (3) winter triticale/double crop forage sorghum-forage sorghum-spring oat rotation with single tillage after spring oat (T/S-S-O RT); (4) winter triticale/double-crop forage sorghum-forage sorghum-forage sorghum-spring oat rotation with no-tillage (T/S-S-S-O NT); (5) winter triticale/double-crop forage sorghum- forage sorghum- forage sorghum-spring oat with single tillage after spring oat (T/S-S-S-O RT); and (6) a winter triticale-forage sorghum-spring oat with no-tillage (T-S-O). There were four replications for each treatment and plot size was 9.1 x 9.1 m.

Data analysis was conducted in SAS 9.4 (SAS Institute). Crop and rotation treatment effects on response variables (forage yield, nutritive value, cost of production, and return) were analyzed using PROC MIXED procedure. For each response variable, a type 3 test of fixed effects was conducted, with each response variable modeled against fixed variable crop or treatment, and year and replication were random effect variables. For a significant ($\alpha = .05$) main effect, a mean separation test was conducted using Tukey's honest significant difference.

Two of the crop rotations in this study, T/S-S-O and T/S-S-S-O, have multiple sorghum crops at different rotation order. For that reason, we conducted a separate analysis to identify if sorghum at different rotation sequences had similar nutritive value and net return. Before analysis, we grouped each of these different sorghum groups as the sorghum in double crop (T/S), and the sorghum after sorghum (-S-S-).

Results and Discussion: Averaged over the years, there was no significant treatment effect on oat forage yield due to tillage and crop rotation treatments. Forage sorghum yield was 26% greater in the S–S compared with T/S–S–O (NT) treatment. Triticale forage yield was 41% more in the T/S-S-S-O (RT) compared with T/S–S-S–O (NT) or T/S-S-O (RT) treatments. Overall, annualized forage yield was greatest for S–S and T/S–S–S–O (NT and RT) and was the least for T/S–S–O (NT) and T–S–O treatments (Table 1).

Nutritive value of the three forage crops in the study varied as expected. Triticale had the greatest CP, forage sorghum had the greatest ADF and NDF, and oat had the greatest Ca, K, P, and IVTDMD compared with the other forage species. The higher nutritive value of oat and triticale combined with the higher yield of forage sorghum integrated into one crop rotation is a benefit of the forage rotation, i.e., to supply all necessary nutrients and biomass yield to meet livestock demand. Crude protein is an essential nutrient for livestock and is positively correlated with both dry matter digestibility and organic matter digestibility where as ADF and NDF are negatively correlated with forage digestibility. Crude protein content decreases with maturity and plants were harvested at heading in this study to optimize both yield and nutritive value. Legumes (that were not included in our study) have much higher CP than grasses, however, in semi-arid regions, legumes tend to produce less dry matter and often not enough to cover the cost of production (Holman et al. 2018, 2022).

Among treatments, there was a 23-31% greater CP for T-S-O and T/S-S-O compared with S-S (Table 2). The differences in CP among rotation treatments is consistent with individual crop CP. The presence of forage sorghum in the rotation tended to decrease CP concentration because forage sorghum had less CP than triticale or oat. The nutritive value differences were minimal among rotation treatments with the exception that S-S tended to have greater ADF and NDF, and lower CP, IVTDMD, Ca, and P concentrations compared with other treatments. Of the treatments, S-S was among the top in net return due to greater forage sorghum yields. However, the T/S-S-S-O (RT) treatment was among the greatest in all four measures, i.e., productivity, stability (based on prior report), nutritive value, and net return (based on this report). Our findings also showed raising forage sorghum and triticale may be more economical compared to some purchased alternatives. We concluded forage producers should consider growing a rotation of forages throughout the year to reduce production variability and weather risk, while increasing forage yield, nutritive value, and market value in designing profitable forage systems.

Literature Cited:

Holman, J., Arnet, K., Dille, A., Kisekka, I., Maxwell, S., Obour, A., Roberts, T., Roozeboom, K., Schlegel, A., 2018. Can cover (or forage) crops replace fallow in the semiarid Central Great Plains? Crop Sci. 58, 1-13.

Holman, J.D., Obour, A. K., Assefa, Y., 2022. Productivity and profitability with fallow replacement forage, grain, and cover crops in W-S-F rotation. Crop science, submitted.

Holman, J. D., Thompson, C. R., Hale, R. L., Schlegel, A. J., 2010. Forage yield and nutritive value of hard red and hard white winter wheat. Agron. J. 102, 759–773.

Holman, J. D., Schlegel, A., Obour, A. K., Assefa, Y., 2020. Dryland cropping system impact on forage accumulation, nutritive value, and rainfall use efficiency. Crop Sci. 60, 3395–3409.

Holman, J. A., Obour, A. K., Assefa, Y., 2021. Rotation and tillage effects on forage cropping systems productivity and resource use. Crop Sci. 61, 3830-3843.

Table 1. Forage accumulation in oat, forage sorghum, and triticale for six rotation by tillage treatments at Garden City, KS, average for years 2013 – 2020.

Treatments	Oat	Forage sorghum	Triticale	Treatment	
				Total	Annualized
	-------------------kg/ha-------------------			kg/ha-cycle	kg/ha
S-S (NT)	-	6932a	-	6932c	6932a
T/S-S-O (NT)	1895	5480b	3107bc	14938b	5171b
T/S-S-O (RT)	1865	5985ab	4261ab	17457b	5966ab
T/S-S-S-O (NT)	1935	6646ab	3113c	24371a	6797a
T/S-S-S-O (RT)	1788	6338ab	4409a	24214a	6958a
T-S-O (NT)	1884	6182ab	3902abc	13169b	4390b
HSD[2]	NS	1072	1165	4231	966
Pr > F	0.896	0.019	0.001	<.0001	<0.0001

[1]Mean values within a column followed by the same letter or with no letter are not significantly different (P<0.05).

[2] HSD is minimum difference between two treatments used to declare they are significantly different using Tukey's Honest Significant Difference Test at p<0.05.

Table 2. Crude protein concentration in oat, forage sorghum, and triticale for six rotation x tillage treatments or double and single crop at Garden City, KS, average for years 2013 – 2020.

Treatments	Oat	Forage sorghum	Triticale	Composite
				Weighted average
	-----------------------------Crude protein (g/kg)-------------------------------------			
S-S (NT)	-	120b[1]	-	113c
T/S-S-O (NT)	167	129a	181	132ab
T/S-S-O (RT)	168	126ab	179	130abc
T/S-S-S-O (NT)	169	124ab	183	125bc
T/S-S-S-O (RT)	165	127ab	178	129abc
T-S-O (NT)	169	126ab	182	139a
HSD[2]	NS	7	NS	14
Pr > F	0.855	0.042	0.615	0.002
Double crop sorghum (T/S)		133a		
Single crop sorghum (S-S)		122b		
HSD		2.87		
Pr > F		<0.001		

[1]Mean values within a column followed by the same letter or with no letter are not significantly different (P<0.05).

[2] HSD is minimum difference between two treatments used to declare they are significantly different using Tukey's Honest Significant Difference Test at p<0.05.

Table 3. Gross return for each crop by treatment and average gross return and net return for each treatment in the forage rotation system study at Garden City, KS.

Treatments	Oat	Forage sorghum	Triticale	Treatment	
				Total for complete rotation	Annualized
	----------------------Gross Return ($/ha)------------------------				Net Return $/ha
S-S (NT)	-	501a	-	501c	50a
T/S-S-O (NT)	207	407b	312bc	1333b	-52b
T/S-S-O (RT)	211	445ab	419ab	1519ab	43a
T/S-S-S-O (NT)	222	463ab	307c	1917a	-12ab
T/S-S-S-O (RT)	206	448ab	441a	1991a	51a
T-S-O (NT)	219	474ab	330abc	1023b	-34b
HSD[2]	NS	75	116	363	88
Pr > F	0.895	0.018	0.001	<0.001	0.001
Double crop sorghum (T/S)		368b			13b
Single crop sorghum (S-S)		491a			60a
HSD		25			17
Pr > F		<0.001			<0.001

CRABGRASS RE-SEEDING AND GRAZING MANAGEMENT DEMONSTRATION

J.D. Huff[1]

Crabgrass is used to provide high-quality, palatable forage for summer grazing. This annual re-seeding forage also provides the reputable re-growth during summer months. The number of seasonal grazing cycles can be enhanced by grazing management strategies. Many producers underestimate the quality and productivity of crabgrass based upon previous assumptions. This demonstration was conducted to provide others a seasonal timeline on strategies to enhance crabgrass emergence, increased forage productivity, and maintain adequate, high-quality forage. The demonstration was conducted on a pasture in western Arkansas that was established in crabgrass in 2019. The variety of crabgrass planted was Mojo, and it has voluntarily re-seeded in 2020 and 2021. The demonstration site is approximately 12-acres of low fertility, acidic silt loam soils. Ten to fifteen cows, heifers and calves solely grazed this site from early May to October. Early grazing was successful by 1) restricting grazing on the site in the late winter and early spring months 2) minimum cool season forage competition, 3) a prescribed burn, and 4) a timely nitrogen application. The cattle were moved daily over 90% of the grazing season. Temporary electric fence and watering tanks provided the framework to flexible grazing management strategies. Stock density ranged from 10,000 to 20,000 lbs./acre. The grazing management strategies provided increased forage availability and days of grazing compared to conventional continuous grazing methods. Narrated video recordings were completed throughout the 2021 grazing season to provide detailed observances, challenges, and successes of growing and grazing crabgrass. This timeline of events will be provided in a video presentation.

[1]Grazing Lands Specialist/Producer, USDA-NRCS, Mena, AR (jeremy.huff@usda.gov).

DRY FERTILIZER IMPREGNATION WITH GRAZONNEXT HL AND DURACOR HERBICIDES: BENEFITS AND LIMITATIONS

S.H. Ingram[1]

Proper soil fertility and weed control are critical components for maintaining and producing quality forage grasses in a grazing or haying system. Of the forage acres that are soil tested, the majority need lime and/or a macronutrient fertilizer (N-P-K) and in many cases an increase in forage yield and nutritive value is observed following a fertilizer application. However, broadleaf weeds, if present, will compete for these applied nutrients and have the potential to negatively affect yield and overall forage consumption. A relatively new concept for forage producers, impregnation of dry fertilizer with herbicide commonly referred to as DFI, can greatly reduce broadleaf weed pressure while also providing the necessary nutrients for optimum forage production. Soil active herbicides such as DuraCor (Rinskor + aminopyralid) and GrazonNext HL (2,4-D + aminopyralid) can be used as herbicide application of these products rely heavily upon rainfall to move the herbicide into the soil profile for plant root uptake. Field trials investigating DFI have shown 85-99% visual weed control for species such as Hairy Buttercup, Curly Dock, and Musk Thistle. Beyond weed control, producers should expect to save time and money by reducing the number of passes across the range, pasture, or hayfield. A DFI application satisfies two key forage management considerations with one application and should be considered a valuable approach to maintaining the most economical feed for livestock.

[1]Integrated Field Scientist, Corteva Agriscience, Savannah, GA (sam.ingram@corteva.com).

IMPACTS OF SMALL GRAIN CEREAL SILAGE MANAGEMENT ON FERMENTATION AND NUTRITIVE VALUE – A SURVEY OF NEBRASKA PRODUCERS

Alexa Johnson, Jenny Rees, Ben Beckman, Brad Schick, Gary Lesoing, Erin Laborie, Kim Clark, Connor Biebler, Daren Redfearn, and Mary Drewnoski[1]

Small cereal silage can be an excellent double crop opportunity. Understanding the current management practices, issues, and questions of producers using small cereals for silage can help generate research and educational material to assist producers in making the best use of this feed resource. In 2021, seventeen Nebraska producers provided twenty small grain silage samples at harvest and again after fermentation to evaluate changes in nutrient content and the quality of the fermentation achieved. A density sample was also taken after fermentation to evaluate effectiveness of packing. Of the samples obtained 55% were cereal rye, 25% triticale, 5% wheatlage, 5% raising oatlage, and 10% mixes. All had raised small cereal silage for three or more years. When surveyed about their production practices, 85% stated they had wilted or field dried the crop, yet 40% of those that wilted still had dry matter content lower than 30%. The average loss in the total digestible nutrients (TDN) from the pre fermentation to after fermentation was 6 TDN units. Samples with greater TDN loss were wet (less than 30% DM) and many had increased butyric acid content, suggesting fermentation by clostridial bacteria. Not only does clostridial fermentation result in loss of energy content but also reduces palatability. The density of silage packed into the silo averaged 4.86 lb DM/ft^3, which is less than a 3rd of the recommend target (15 lb DM/ft^3). These data suggest that moisture management and packing are a challenge for producers and an area for improvement that could help improve the quality of their small grain silage.

[1]University of Nebraska.

REACHING GEORGIA FORAGE PRODUCERS IN A VIRTUAL ENVIRONMENT

C.H. Knight[1], L. Baxter[2], P. Brown[3], R. Gates, J. Kichler, S. Morgan, L. Ray, S. Tanner, P. Sapp[4]

COVID-19 continues to create uncertainty and concern both globally and locally. Lockdowns interrupted conventional Extension programming nationwide. The University of Georgia Forage Extension Team and Forage Extension Specialist maintained learning during this period through virtual programs and webinars. The team also reached clientele through social media outlets including YouTube, Facebook, Twitter, and Instagram. In the Fall, 2020, the GrassMasters Course was offered virtually through the Zoom platform. Other virtual programming included Dairy Forage Field Day, Georgia Forages Conference, Hay and Baleage Short Course, and Corn Silage and Stored Forage Field Day. The quarterly GrassMasters Spotlight Series highlighted topics identified on past GrassMasters evaluations as subjects where participants wanted more information. These are open to any registrant via Zoom, and recorded and posted to Georgia Forages YouTube Channel. The UGA Forage Extension Specialist created and produced (utilizing WeVideo) Georgia Forages: The Digital Series, a video series posted on YouTube and hosted on Facebook. Infographics (created in Canva) are posted in Twitter, Instagram, and Facebook to highlight data and troubleshooting information. The UGA Forage Extension Team Newsletter is emailed and posted to the Forage Team Blog. The UGA Forage Extension Specialist hosts monthly in-service trainings via Zoom with Agents to discuss research and forage production issues. Although not ideal, virtual programs have been well received and allowed programming to continue and expand in a challenging situation.

[1]Ag and Natural Resources Agent, Univ. of Georgia (UGA) Extension, Madison County, Danielsville, GA (clh@uga.edu); [2]Extension Forage Specialist, UGA Crop and Soil Science, Tifton, GA (baxterl@uga.edu), [3]Grasslands Conservationist, Georgia NRCS, Athens, GA 30602; [4]Ag and Natural Resources Agents, UGA Extension, UGA Forage Extension Team.

IMPLEMENTING RESOURCE CONSERVATION PRACTICES ON HORSE FARMS

Krista L. Lea[1] and S. Ray Smith[2]

Conservation practices on horse farms have lagged compared to cattle operations, likely due to hesitancy on the part of both the farm owner and NRCS conservationist. This Regional Conservation Partnership Program (RCPP) served as a bridge between these two parties to implement resource conservation practices on horse farms, as well as demonstrate their feasibility on other farms, and to educate conservation planners. Twenty farms participated in the program, installing fencing, automatic waterers, heavy use areas and forage and biomass plantings, among other CAP 110 practices. The farms served as demonstration farms and hosted educational field days to show other horse owners how they used technical and financial assistance from NRCS to improve their operations and improve the environment. Two trainings for NRCS conservationists were also held to introduce them to horse practices and how NRCS standards can be applied to horse farms. Finally, practice standards that were insufficient for horses, (e.g.-square feet allowed for heavy use areas) were changed to better meet the needs of the animals, their managers, and provide better protection for the environment. Work from this program is continuing with a NRSC - Conservation Innovation Grant to study the environmental and economic impacts of improving horse pastures.

[1]Research Analyst and [2]Extension Professor. University of Kentucky, Department of Plant and Soil Sciences, Lexington KY (Krista.Lea1@uky.edu).

WARM-SEASON ANNUAL GRASSES AS COMPANION FORAGE CROP FOR ESTABLISHING WARM-SEASON PERENNIAL GRASSES

R. Lemus[1] and J. White[2]

Bahiagrass (BH, *Paspalum notatum*) and bermudagrass (BG, *Cynodon dactylon*) are warm-season perennial grasses (WSPG) commonly grown in the southern United States. The objective of the study was to determine the establishment of WSPG using warm-season annual forages as a nursery crop. The study was a split-plot design replicated three times. The main plots were BH, and BG established at seeding rates of 20 and 10 lb PLS/ac. The subplots were three-warm-season annual grasses (WSAG) ['AF7401'forage sorghum (FS), 'Prime 180' pearl millet (PM), 'Green Grazer V' sorghum-sudan hybrid (SS)], established at 10, 15, and 20 lb PLS/ac. All treatments received 100 lb N/ac in split applications after WSAG germination and following the first WSAG harvest. Treatments were harvested using a Wintersteiger Cibus F harvester (Wintersteiger AG, Austria) when plants reached 48 in height. Biomass subsamples were analyzed for nutritive value using a Foss DS2500 NIR instrument (Foss North America, Eden Prairie, MN) and the 2021 mixed grass hay equation from the NIRS Forage and Feed Testing Consortium (Berea, KY). Data were analyzed in the PROC GLIMMIX of SAS at $\alpha = 0.05$. Bahiagrass and bermudagrass establishment and productivity were not affected by WSAG. Total biomass production was 18% greater for the BH treatments compared to BH. The yield index of adding FS, PM, and SS were 108, 38, and 25% when compared to WSPG monocultures. At first harvest, CP, WSC, and IVTDMD concentrations were not different among WSAG treatments.

[1]Professor and Extension Forage Specialist, Mississippi State University, MS (Rocky.Lemus@msstate.edu); [2]Forage Variety Testing Manager, Mississippi State University, MS (JWhite@pss.msstate.edu).

FORAGE PRODUCTIVITY, WEED DENSITY, AND SOIL PROPERTIES AS AFFECTED BY GRAZING AND TILLAGE OF ANNUAL FORAGES

P.S. Mauler, J.D. Holman, L.M. Simon, A.K. Obour, S.K. Johnson, and K.L. Roozeboom[1]

Abstract

No-till (NT) management of annual crop production systems increases precipitation capture and storage. However, potential soil compaction with annual livestock grazing and the development of herbicide resistant weeds present challenges to long-term NT. An on-farm study was established in 2016 near Jetmore, KS to investigate the effects of annual minimal tillage (MT) with a sweep plow compared to NT in a grazed (averaged 5 cow-calf pairs ha^{-1}) continuous winter triticale production system. Forage productivity as well as weed density and soil properties were determined in the 2020 and 2021 production years before and after the implementation of summer tillage. Soil properties were compared to an adjacent perennial grass pasture in 2020. Results showed that MT plots had 4.5-fold lower weed density compared to NT in summer 2021 before tillage occurred. Additionally, soil chemical and physical properties in 2020 were unaffected by tillage compared to NT in the 0- to 5-cm and 5- to 15-cm soil depths. However, when compared to an adjacent perennial pasture, both tilled and NT plots had greater soil bulk density (13%), lower soil organic carbon (40%), and lower mean weight diameter of water stable aggregates (69%). These preliminary results suggest that MT of grazed annual croplands reduces weed density without affecting soil properties.

Introduction: Growing annual forages in semiarid dryland cropping systems in the central Great Plains provide many benefits for livestock production and can be grazed to either supplement or rest native rangeland. Triticale is a winter hardy annual forage that has greater productivity and comparable nutritive value compared with wheat. Winter triticale is usually planted in the fall and grazed over the winter, then either grazed out in the spring- or harvested- as a hay or silage crop. Grazing annual forages can be more profitable than haying because it eliminates mechanical harvesting and feed delivery expense. No-till (NT) management of annual forage production systems increases precipitation capture and storage and improves soil health. However, compaction from livestock and development of herbicide resistant weeds present new challenges to no-till production. Annual tillage with a sweep plow can help breakup compaction from annual grazing as well as reduce weed density without affecting soil properties. The objective of this experiment was to quantify the effect of annual tillage compared to NT management on annual forage productions, weed density and soil properties in continuous triticale production systems.

---

[1] Graduate Research Assistant, Southwest Research-Extension Center, Kansas State University, Garden City, KS; Professor, Southwest Research-Extension Center, Kansas State University, Garden City, KS; Graduate Research Assistant, Agricultural Research Center-Hays, Kansas State University, Hays, KS; Associate Professor, Agricultural Research Center-Hays, Kansas State University, Hays, KS; Professor, Northwest Research-Extension Center, Kansas State University, Colby, KS; Professor, Department of Agronomy, Kansas State University, Manhattan, KS.

Materials and Methods: An annual forage grazing experiment was initiated in 2016 at an on-farm research field near Jetmore, KS. The study was a randomized complete block design with 4 replications. The two tillage treatments were NT and MT implemented in a grazed continuous winter triticale cropping system. Plots were 50ft wide and 1300 ft long.

Tillage was implemented twice during the fallow period between winter triticale crops using a sweep plow ('Noble plow') sometime between 1 July and 1 August. The sweep plow is a minimum disturbance equipment commonly used in the region for weed control. All plots received herbicides for controlling weeds.

Triticale was planted between 15 August and 15 September and was grazed during the winter and spring. Stocking rate were 500 lbs./ac during the fall and 1000 to 2000 lbs./ac in the spring. Livestock was removed to an adjacent native grass pasture either before or soon after heavy rain events to minimize surface compaction. Grazing ended between 15 May and 15 June once the triticale headed or was grazed out, depending on the year.

Forage yield in the spring and summer and quality measured by using two 2.5ft^2 quadrants randomly placed in each plot. Weed density was measured pre-tillage in June and post-tillage in August by using a 100 ft^2 quadrats. Soil bulk density, soil organic carbon, and water stable aggregates were measured by collecting soil samples at two depth increments of 0 to 2 inches to 2 to 6 inches as these two periods. Statistical analyses were done using PROC GLIMMIX of SAS ver.9.4 (SAS Institute, 2012, Cary, NC) with year (or sampling time) and treatment considered fixed and replication considered random. Treatment differences were considered significant at $P \leq 0.05$.

Results and Conclusions: Available forage production measured in March was greater with MT but forage mass was unaffected by tillage in June (Fig. 1). Weed density was different between tillage treatments. For example, both June and August of 2020, weed density with Mt was significantly less compared to NT. In June 2021, MT treatments had a 4.5 times lower weed density compared to NT (Fig. 2). Bulk density was not different between MT and NT at both 0-2 inch and 2–6-inch depths. Water stable aggregates showed no difference in 2020 in both June and August. In 2021, mean weight diameter (MWD) was similar in June however, in August, MWD decreased with MT due to greater tillage intensity compared to 2020 (Fig. 3).

In conclusion, MT had no negative effect on forage production- compared with NT. Minimal tilled plots had 450% less weed density compared to NT. The preliminary results showed that MT of grazed winter triticale reduces weed density and can increase early available forage without negatively affecting soil chemical and physical properties

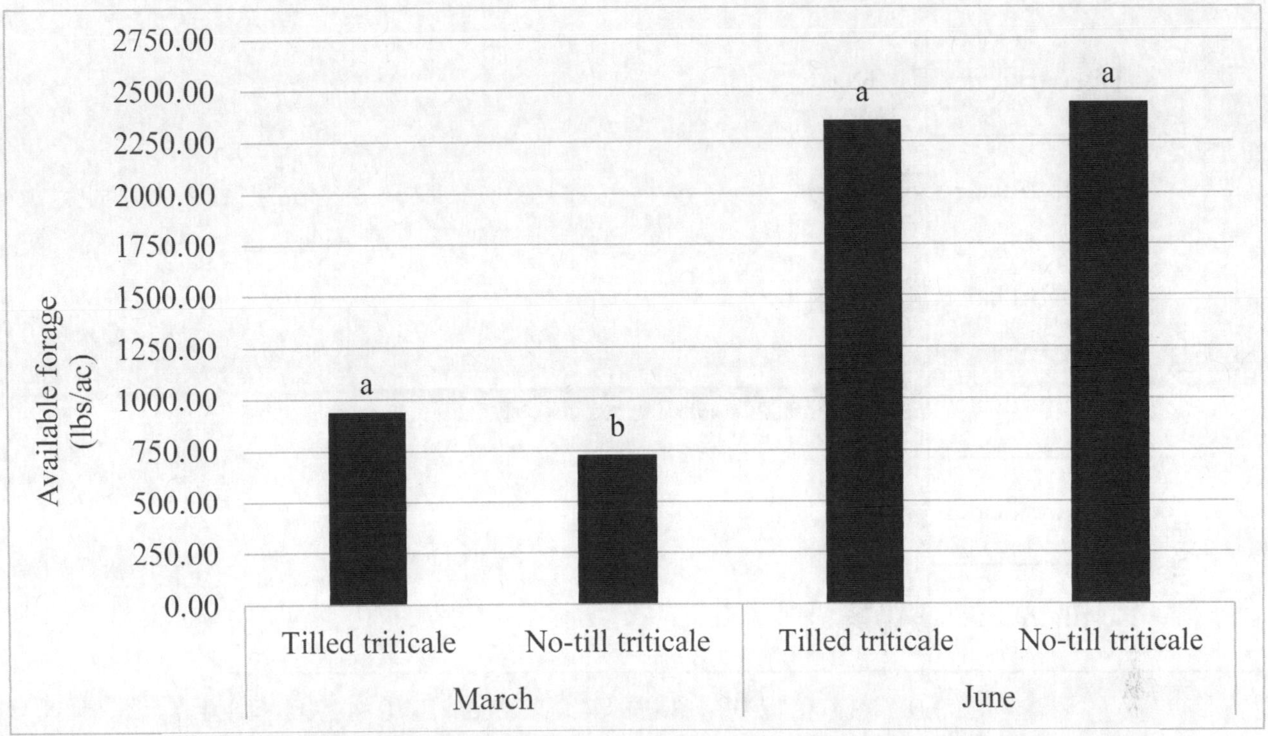

Figure 1. Tillage effects on triticale forage production in 2021 near Jetmore, KS.

[†]Means with the same letter are not significantly different ($P < 0.05$) among years.

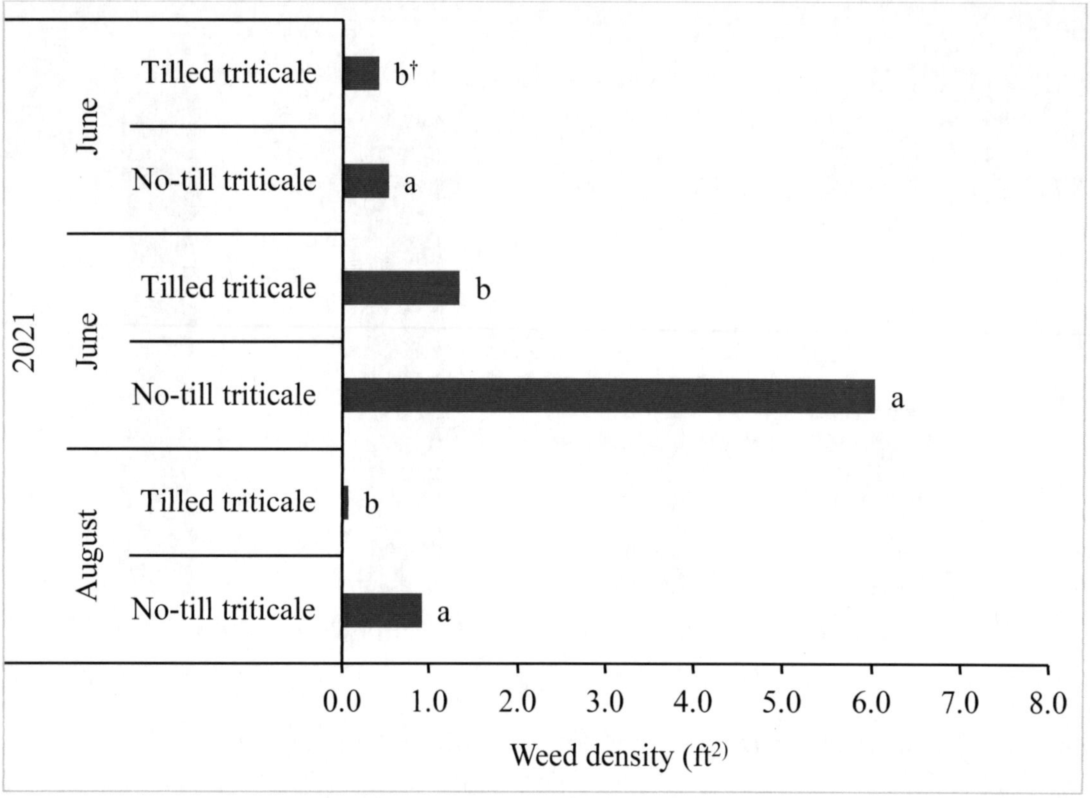

Figure 2. Tillage effects on weed density in a grazed winter triticale system in 2020 and 2021 near Jetmore, KS.

†Means with the same letter are not significantly different (P <0.05) among years.

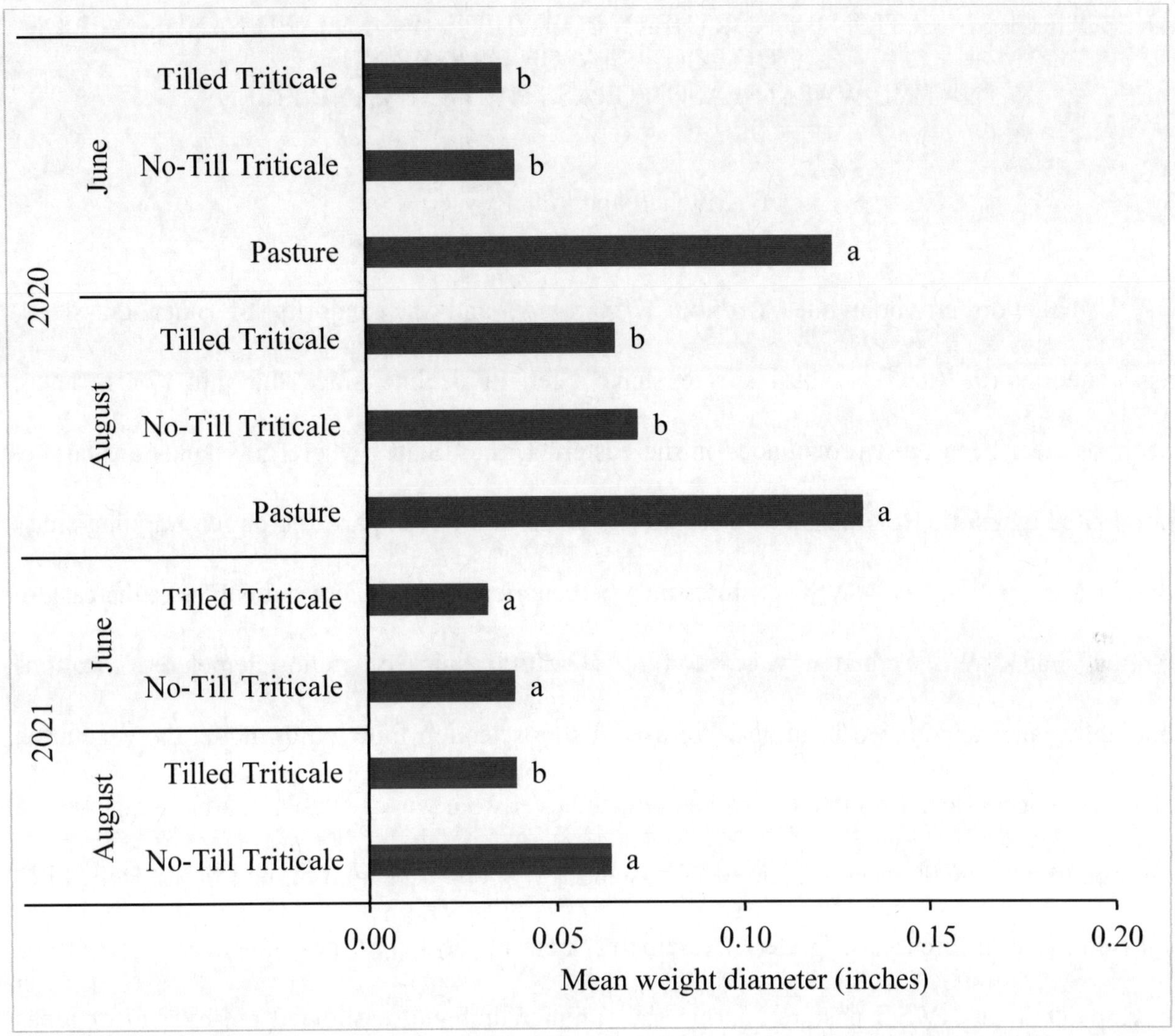

Figure 3. Tillage effects on mean weight diameter of water stable aggregates in the 0-5 cm soil depth in a grazed winter triticale system in 2020 and 2021 near Jetmore, KS.

†Means with the same letter are not significantly different (P <0.05) among years.

RESOURCE SELECTION BY NORTHERN BOBWHITES IN A MIXED LANDSCAPE OF EXOTIC COOL-SEASON GRASS AND NATIVE-WARM SEASON GRASS PASTURES AND FIELDS

D. Mitchell and P.D. Keyser[1]

The northern bobwhite (*Colinus virginianus*) and the majority of other grassland-associated North American bird species have been in decline since the mid-1960s. These declines have been most pronounced in the eastern United States where grasslands are largely dominated by exotic CSG pastures. This study was carried out as part of a larger, ongoing study that is investigating how NWSG pastures may be beneficial for northern bobwhites in the eastern United States. We evaluated vegetation at over 700 pairs of radio-telemetered locations bobwhites and randomized locations. We used a step-selection function to model the vegetation characteristics of locations that bobwhites selected over what was available in a landscape with a mixture of exotic CSG pastures and hayfields and NWSG pastures and burned fields. Bobwhites showed some differences in preferences among some of the field types, however, some trends were evident. In NWSG pastures during the summer, bobwhites showed preference for areas with more bare ground and avoided areas with increased CSGs and litter (CoxPH coeff, $p \leq 0.05$; +1.27, -0.95, and -3.91 respectively). In all field types, wintering bobwhites showed preference for areas with a higher percentage of woody stems than would be expected at random (CoxPH coeff, $p \leq 0.05$; range +2.05 to +3.79). The results of our study highlight the preferences of northern bobwhites in a working-lands context. Property owners may be able to apply this information to improve numbers of bobwhites in their cattle or forage production operations.

[1]Graduate Research Assistant, UT, Knoxville, TN (dmitch41@vols.utk.edu); Professor and Director, CNGM, UT, Knoxville, TN (pkeyser@utk.edu).

MIDWEST COVER CROPS COUNCIL,
EMPOWERING A NETWORK OF PROFESSIONALS

Anna Morrow[1]

The mission of the Midwest Cover Crops Council (MCCC) is to facilitate widespread adoption of cover crops across the Midwest. We execute this mission through education, outreach materials, and networking. The MCCC is a clearinghouse for cover crop resources. Our website (midwestcovercrops.org) hosts extension materials including cover crop recipes, intended to provide a low risk strategy for farmers new to cover cropping. The popular cover crop pocket field guide has been revised and now includes additional species and expanded management sections. The online cover crop selector tool uses county level climate data to aid in seeding date and species selection based on goals. With the growth of cover crop adoption and increased interest integration of livestock in these systems, MCCC is excited to expand our network to include more grazing specialists and resources. The overlap between forages and cover crops makes this a natural fit. Anna will discuss materials and upcoming resources from MCCC including updates for Kansas and other states, new calculators, crop advisor modules, and new opportunities. Join MCCC in reaching new audiences interested in forages, grazing, and cover crops.

[1]Midwest Cover Crops Council (annamorrow@purdue.edu).

SPRING OAT AND TRITICALE FORAGE PRODUCTION AS AFFECTED BY SEEDING RATE AND NITROGEN FERTILIZER APPLICATION

A.K. Obour[1], J.D. Holman[2], L.M. Simon[1], and P.S. Mauler[1]

Abstract

Growing spring forages could reduce soil erosion and increase cropping system profitability compared to summer-fallow in a wheat (*Triticum aestivum* L)–sorghum (*Sorghum bicolor* L. Moench ssp. *Bicolor*)-fallow system. Spring oat (*Avena sativa* L.) and triticale (*X Triticosecale* Wittmack) are important cool-season annual forage crops in dryland production systems. However, information on best agronomic management practices including seeding rate and nitrogen (N) fertilizer recommendations are limited. Field experiments were conducted to evaluate seeding rate and nitrogen (N) application effect on oat and triticale forage production. Treatments were split-plot arrangement of three seeding rates (75, 100, and 125% of recommended seeding rate) as main plots and five N rates (0, 10, 30, 50, and 70 lb N/a) as subplot. Results showed seeding rate had no effect on forage mass or nutritive value in both oat and triticale. Across sites, oat forage mass produced with 30 lb N/a was similar to yields with greater N rates. Similarly, triticale forage productivity increased with N application rates but not beyond 50 lb/a. Increasing N application rates increased crude protein (CP) and *in-vitro* dry matter digestibility (IVDMD), while decreasing acid detergent fiber (ADF) and neutral detergent fiber (NDF) concentrations. Our findings suggest growers can plant oat and triticale forage at 25% less the recommended seeding rate of 64 lb/a for oat and 72 lb/a for triticale with no in forage mass.

<u>Introduction:</u> Spring oat and triticale are important forage crops in dryland and limited irrigated crop production systems in western Kansas. Previous research showed growing spring forages in place of fallow reduced soil erosion and increased fallow precipitation use efficiency compared to summer fallow (Blanco et al., 2013), and increased profitability compared to fallow in years with average to above average rainfall (Holman et al., 2018, 2020). Despite the great potential of spring forages, information on seeding rate and nitrogen fertilizer recommendations are limited. These management practices can affect forage productivity and nutritive value.

Current N fertilizer recommendations for oat and triticale are based on a very limited dataset. Residual nutrients from the previous crop could provide adequate nutrients for the spring forage crop. However, lower spring temperatures and N immobilization from the previous crop residue often limit early-season N availability for oat or triticale forage. It is therefore imperative that site-specific N fertilizer research are conducted to fine tune N fertilizer rates for these cool-season forages to improve yields and environmental quality.

---

[1]Agricultural Research Center-Hays, Kansas State University, Hays, KS; [2]Southwest Research-Extension Center, Kansas State University, Garden City, KS.

Seeding rate is an important crop management practice that affects forage production. Spring oat and triticale grown for forage are planted at 25 to 50% greater seeding rates than when managed for grain production. This increase in plant density will allow for greater early crop establishment, smaller stems, and increased forage production. Determining the optimum seeding rate for oat and triticale is important because seed costs constitute a significant component of the variable cost in forage production systems. The objectives of this study were to (1) determine N fertilizer rate effects on forage production and nutritive value for oat and triticale forage, and (2) quantify effect of seeding rate on oat and triticale forage yield and its interaction with N fertilizer application under dryland conditions in western Kansas.

Materials and Methods: Field experiments were conducted over four site-years across three locations (Garden City, Hays, and Jetmore) in western Kansas during 2015 and 2016 growing seasons. The experimental design was a randomized complete block with three replications in a split-plot arrangement. The main plots were three seeding rates (75, 100, and 125% of the recommended oat and triticale forage seeding rates) and five N fertilizer rates (0, 10, 30, 50, and 70 lb N/a) as sub-plots. The 100% oat forage recommended seeding rate was 64 lb/a and the 100% recommended seeding rate for triticale was 72 lb/a. Individual plots sizes were 10 ft. wide by 30 ft long. Nitrogen source at all site-years was urea, which was broadcasted after crop emergence. Plots at each site-year received a blanket application of 15 lb P_2O_5/a prior to seeding. Weed control was accomplished with a pre-plant burn down of glyphosate [isopropylamine salt of *N*-(phosphonomethyl) glycine] and 2, 4-dichlorophenoxyacetic acid. Planting for both crops at each site-year were done in March, with planting date dependent on weather and soil conditions.

Oat and triticale were harvested at heading to optimize forage DM accumulation and nutritive value. Forage harvests were performed the first or second week in June. During each harvest, a 3 ft. × 30 ft. forage strip was harvested from each plot using a Carter plot forage harvester (Carter Manufacturing Company, Inc.) to 3 inch stubble height. Fresh weights of samples were recorded, sub-samples were weighed, and oven dried at 60°C for at least 48 hours in a forced-air oven for DM determination. Oven-dried samples were ground to pass through a 1-mm mesh screen in a Wiley Mill (Thomas Scientific, Swedesboro, NJ). The ground samples were then analyzed for forage nutritive value [crude protein (CP), acid detergent fiber (ADF), neutral detergent fiber (NDF), *in-vitro* dry matter digestibility (IVDMD)], and tissue nutrient concentrations (Ward Laboratories, Inc., Kearney, NE) using Foss 6500 near infrared spectroscopy (NIRS).

Statistical analyses were completed using PROC GLIMMIX of SAS ver. 9.4 (SAS Institute, 2012, Cary, NC) with seeding rate and N rates considered as fixed effects whiles block nested within location were treated as random effects in the model. Treatments and their interactions effect were significant when F test P values were ≤ 0.05.

Results and Discussion: Results of this study showed seeding rate had no significant effect on oat ($P = 0.80$) or triticale ($P = 0.59$) forage dry matter produced. Averaged across site-years and N fertilizer rates, oat forage mass was 3592 lb/a with 75% of the recommended seeding rate, 3679 lb/a for 100% recommended seeding rate and 3616 lb/a for the 125% of the recommended seeding rate treatment. Similarly, triticale forage mass was 4106 lb/a with 75% of th e recommended seeding rate, 4362 lb/a for 100% recommended seeding rate and 4280 lb/a for 125% recommended seeding rate. This finding suggests the current recommended seeding rate for spring oat and triticale managed for forage, 64 and 72 lb/a, respectively, could be reduced 25% without significant yield penalties. This is important because seed costs account for a significant portion of the variable cost in forage production. Savings accrued through reduced seeding rate could increase profitability of the production system.

Application of N fertilizer has a significant effect on both oat and triticale forage production. Oat forage mass with 30 lb N/a was not different from that obtained with greater N fertilizer rates (Table 1). Similarly, across site-years, triticale forae mass of the unfertilized control was 80% of that obtained with 70 lb/a. However, triticale forage plateaued with 50 lb N/a. In general, CP concentration increased with an increase in N fertilizer application (Tables 1). Averaged across site-years, N fertilizer application to oat increased CP concentration from 11.7 % with the unfertilized control to 13.8 % when 70 lb N/ was applied. Crude protein concentration in triticale forage ranged from 12.8% with no fertilizer N to 14% when 70 lb N /a was applied. The CP requirement for growing replacement heifers with body weight (BW) of 1200 lb at maturity ranged from 10.2% (with growing BW of 660 lb) to 8.1% (with growing BW of 960 lb) assuming the forage contains ≥ 60% total digestible nutrients (NRC, 2000. Nutrient requirement of beef cattle). Therefore, average CP concentration of the treatments in the current study were greater than the minimum CP requirement for growth or maintenance of grazing beef cattle.

Oat and triticale ADF, NDF, IVDMD and nutrient concentration were affected by N fertilizer application (Table 2). In general, ADF and NDF concentrations decreased with increase in N fertilizer application rate in both oat and triticale forage. Lignin concentration decreased slightly with increased N fertilizer rate (Table 2) and could partly account for the observed decrease in ADF and NDF concentrations with N fertilizer application. The IVDMD concentration as a measure of forage digestibility increased with N fertilizer application rate in both oat and triticale. This observation was expected because application of N fertilizer decreased ADF and NDF concentrations, both of which affect forage digestibility. Forage oat calcium (Ca), phosphorus (P) and potassium (K) concentrations increased with an increase in N fertilizer application rate. However, P concentration in triticale forage was not affected by N fertilizer application rate. In conclusion, findings of this study suggest spring oat and triticale forage can be planted at 25% less the recommended seeding rate with no decrease in forage yield and provides nutritive values sufficient for growing and maintaining beef cattle.

Literature cited:

Blanco-Canqui, H., J.D. Holman, A.J. Schlegel, J. Tatarko, & T.M. Shaver. 2013. Replacing fallow with cover crops in a semiarid soil: Effects on soil properties. Soil Sci. Soc. Amer. J. 77: 1026-1034.

Holman, J.D., K. Arnet, J. Dille, S. Maxwell, A. Obour, T. Roberts, K. Roozeboom, and A. Schlegel. 2018. Can cover or forage crops replace fallow in the semiarid central Great Plains? Crop Sci. 58:932-944
SAS Institute. 2012. SAS/STAT user's guide, Version 9.3. SAS Inst., Cary, NC.

Table 1. Oat forage mass and crude protein (CP) concentration as affected by N application at three locations in western Kansas

	Oat	Triticale	Oat	triticale
lb/a	Forage mass (lb/a)		Crude protein (%)	
0	2891 c[a]	3721 d	11.7 d	12.8 b
10	3246 b	4021 c	12.0 d	13.1 b
30	3906 a	4269 b	12.5 c	14.0 a
50	4018 a	4548 a	13.2 b	13.6 ab
70	4085 a	4686 a	13.8 a	14.0 a

[a]Means followed by same lower case letter (s) within a column are not significantly different (*P* > 0.05).

Table 2. Oat and triticale acid detergent fiber (ADF), neutral detergent fiber (NDF), *in-vitro* dry matter digestible (IVDMD), lignin, calcium (Ca), phosphorus (P), and potassium concentrations as affected by N application at three locations in western Kansas

Oat	ADF	NDF	IVDMD	Lignin	Ca	P	K
N rate lb/a				%			
0	37.3 a[a]	61.0 a	76.1 c	2.8 a	0.40 b	0.29 c	2.52 d
10	36.8 ab	60.0 b	76.6 bc	2.7 ab	0.40 b	0.29 c	2.61 c
30	36.3 bc	60.0 b	77.2 ab	2.7 ab	0.42 a	0.31 b	2.78 b
50	36.2 bc	59.0 c	77.7 a	2.6 b	0.43 a	0.32 a	2.89 a
70	36.0 c	60.0 b	77.9 a	2.6 b	0.43 a	0.32 a	2.96 a
Triticale							
N rate (lb/a)				%			
0	37.7 a	63.5 a	72.0 b	3.7 a	0.37 b	0.29 a	2.32 c
10	37.6 a	63.6 a	72.1 b	3.7 a	0.38 ab	0.29 a	2.37 c
30	36.8 b	62.5 b	73.3 a	3.5 a	0.38 ab	0.30 a	2.47 b
50	37.1 ab	63.1 ab	72.5b	3.6 a	0.38 ab	0.30 a	2.43 b
70	37.0 ab	62.4 b	73.3 a	3.5 a	0.39 a	0.30 a	2.54 a

[a]Means followed by same letter (s) within a column are not significantly different ($P > 0.05$).

NITROGEN APPLICATION EFFECTS ON FORAGE SORGHUM PRODUCTION AND NITRATE CONCENTRATION

J.D. Holman[2], A.K. Obour[1], L.M. Simon[1], and P.S. Mauler[1]

Abstract

Forage sorghum (*Sorghum bicolor* (L.) Moench) is an important annual forage crop but prone to high nitrate concentration, which can cause toxicity when fed to cattle (*Bos taurus* and *Bos indicus*). Field experiments were conducted over six site-years across Kansas to determine economic optimum nitrogen (N) rate for no-tillage forage sorghum and investigate the effect of N fertilization on sorghum forage nitrate concentration. Treatments were N fertilizer application rates 0, 25, 50, 75, 100 lb N/acre applied prior to planting in 2011 and 2012 near Randolph, KS or 0, 25, 50, 75, 100 and 125 lb N/acre near Garden City (2012 and 2013 growing seasons) and Spearville (2012 and 2014 growing seasons). A quadratic model described the relationship between sorghum dry matter (DM) produced and N rate across the combined site-years. Maximum DM yield of 5855 lb/acre was produced with N application rate of 93 lb N/acre. However, the economic optimum N rate ranged from 0 to 70 lb N/acre depending on sorghum hay price and N fertilizer costs. Crude protein concentration increased with N fertilizer application but N rates beyond 65 lb N/acre resulted in forage nitrate concentrations greater than safe limit of 3000 ppm. Nitrogen uptake increased with N fertilizer application but nitrogen use efficiency (NUE) decreased with N fertilizer rates beyond 50 lb N/acre. In conclusion, forage sorghum required 50 to 65 lb N/acre to produce economic optimum DM yields with safe nitrate concentration.

Introduction: Forage sorghum (*Sorghum bicolor* (L.) Moench) is a highly productive annual summer forage that can be grazed, hayed or ensiled for winter feed, confined feeding, or grazed to supplement native pasture. Recently, there has been interest in growing forage sorghum for hay or grazing in place of fallow to increase residue cover as part of an integrated forage or cover crop system. The high water use efficiency and heat and drought tolerance of sorghum makes it an ideal forage crop choice for growers in the water-limited Great Plains environment. Furthermore, due to the declining saturated thickness of the Ogallala and High Plains aquifers, there is a need for more water-efficient forages such as sorghum compared to corn silage or alfalfa.

Despite the great potential of forage sorghum as animal feed, high nitrate content is a major concern when fed to cattle. Forage containing 3,000 ppm nitrates are generally considered safe as livestock feed, nitrate concentrations between 3,000 and 6,000 ppm may be limit fed, and concentrations > 9,000 ppm should not be fed to livestock. Ensiling reduces nitrate concentration

---

[1]Agricultural Research Center-Hays, Kansas State University, Hays, KS; [2]Southwest Research-Extension Center, Kansas State University, Garden City, KS.

but still requires caution when fed. Environmental stressors (including drought, frost, or cloudy weather conditions) and high manure or N fertilizer application rates can predispose forage crops to accumulate greater nitrate concentration in the aboveground biomass. Nitrogen application generally increased DM production, but excessive N application will also increase nitrate concentration in the harvested biomass. Little research has been conducted on forage sorghum DM production and nitrate concentration responses to N fertilizer application in the semi-arid Great Plains region. Objectives of this study were to determine optimum N rates and quantify forage sorghum hay nitrate concentration as affected by N application under rain-fed no-till conditions in the semiarid central Great Plains.

Materials and Methods: Two field experiments with different N rates were conducted in northeast Kansas near Randolph (2011 and 2012 growing seasons) and in southwest Kansas near Garden City (2012 and 2013 growing seasons) and Spearville (2012 and 2014 growing seasons). In the first study, N fertilizer application rates were 0, 25, 50, 75 and 100 lb N/acre applied prior to planting in 2011 and 2012 near Randolph, KS. The N fertilizer rates in the second study were 0, 25, 50, 75, 100 and 125 lb N/acre applied prior to planting, conducted over four site-years near Garden City and Spearville. At each site-year, the N fertilizer rates were arranged in a randomized complete block design with four replications. Individual plot sizes were 10 ft. wide by 30 ft. long. Weed control at each site-year was accomplished with a pre-plant burn down using appropriate herbicides. Forage sorghum at each site-year was planted the first week in June depending on weather and soil conditions. All plots at each site-year received a blanket application of P and S fertilizers prior to seeding based on Kansas State University soil test recommendations.

Forage plots were harvested at heading which is when most producers harvest to optimize both forge biomass accumulation and nutritive value. Two forage harvests were performed in Randolph, the first harvest in August at heading, and a second harvest occurred the first week in November of regrowth prior to a killing freeze. Due to limited moisture availability to generate regrowth, forage at Garden City and Spearville were harvested only once in August at heading. During each harvest, a 3 ft. × 30 ft. forage strip was harvested from each plot to a 6-inch stubble height using a Carter small plot forage harvester (Carter Manufacturing Company, Inc., Grand Haven, MI). Forage plots were intentionally harvested at a higher stubble height to match Kansas State University recommendations to retain soil residue cover and avoid the lowest portions of the plant, which are normally greater in nitrate concentration. Fresh weights of samples were recorded, sub-samples were weighed, and oven dried at 50°C for at least 48 hours in a forced-air oven for DM determination. Oven-dried samples were ground to pass through a 1-mm mesh screen in a Wiley Mill (Thomas Scientific, Swedesboro, NJ), sequentially. The ground samples were then analyzed for N concentration by dry combustion using a LECO CN analyzer (LECO Corporation, St. Joseph, MI). In addition, forage samples were extracted with 2 M KCl and nitrate concentration determined colorimetrically by cadmium reduction.

Statistical analyses were completed using PROC MIXED procedure in SAS ver. 9.4 (SAS Institute, 2012, Cary, NC) examine sorghum DM production, CP, nitrate concentration, and NUE as a function of N application for the two experiments. Total DM produced, N uptake, and NUE in the first study were analyzed with N rate as fixed effect, year and replication were treated as random effects. In the second study, N rates were considered fixed effects, and replicate nested within site-year as random. A quadratic response model was used to describe the combined DM response to N fertilizer application rate, and the agronomic maximum N rate was determined by equating the first derivative of the quadratic function to zero. Similarly, the economic optimum N rates (EONR) were calculated by setting the first derivative of the quadratic response curve of the relationship between DM production and available N rate equal to the price ratio of the N fertilizer cost and the price of sorghum hay. The EONR rates were then computed for fertilizer price range of $300 to $800/ton of urea fertilizer, and hay value of $40 to $120/ton DM (USDA-AMS, 2018).

<u>Results and Discussion:</u> Nitrogen application had a positive effect on sorghum forage DM production over the six site-years. In eastern Kansas, total forage produced with 100 lb N/acre was 1.6-fold greater than that obtained with the unfertilized N treatment (Table 1). Similarly, in western Kansas, forage sorghum DM production increased linearly with an increase in N application rate. However, DM yields were similar at N rates of 50 lb N/acre or greater (Table 2). Total annual DM yield over the six site-years (both eastern and western Kansas) were combined and modelled and the N rate effect was expressed by a quadratic response function: Yield = 3953 + 41x -0.22x^2, R^2 = 0.75. The model showed maximum forage sorghum DM yield of 5855 lb/acre that occurred with an N rate of 93 lb/acre (~90 lb N/acre). However, the EONR ranged from 0 to 70 lb N/acre depending on N fertilizer costs and forage sorghum hay price.

Application of N fertilizer increased forage sorghum CP concentration. Averaged across harvests, CP concentration increased from 7.1% with the control to 8.6% with 100 lb N/acre in the first study. Likewise, CP concentration ranged from 8.2% to 9.0% for the same N rates for the second study. Forage nitrate concentration in the first study was significantly affected by N rate × harvest timing (Tables 1). Irrespective of N rate, nitrate concentration measured at first harvest was significantly greater than that measured in the regrowth. Averaged across the 2-yr, applying N fertilizer increased forage nitrate concentration in excess of 3000 ppm when N was applied at 75 or 100 lb N/acre (Table 1). Nitrogen application had no effect on nitrate concentration measured in the harvested regrowth, possibly because no additional N fertilizer was applied between forage cuttings that limited the amount of available N for uptake. Across site-years, forage nitrate concentration ranged from 1675 ppm with the unfertilized control to 3551 ppm when 125 lb N/acre was applied in the second study (Table 2). Using the combined forage nitrate data across locations, there was a linear response of forage nitrate concentration to N fertilizer rate. The regression analysis showed a 32 ppm increase in forage nitrate concentration with each lb of N applied (Fig. 1). In general, N application rates beyond 65 lb N /acre resulted in forage nitrate concentration exceeding the safe limit of 3000 ppm for cattle consumption (Roozeboom et al., 2011).

In general, N uptake increased with an increase in N fertilizer application rate. However, NUE computed as DM accumulated per unit of N uptake decreased with increasing N application rate. Applying N fertilizer decreased NUE from 91 lb DM/lb N with the unfertilized control to 75 lb DM/lb N at the highest N rate of 100 lb N/acre in the first study (Table 1). In contrast, N application rate had no significant ($P > 0.05$) effect on NUE in the second study (Table 2), partly due to greater residual soil N concentration at some of the locations in the second study. Our findings showed forage sorghum required 50 to 65 lb/acre to produce economic optimum DM yields without exceeding the safe limit of 3000 ppm for beef cattle.

Literature cited:

Roozeboom, K., D. Blasi, and D.Mengel. 2011: Nitrate Toxicity. Kansas State Univ. Coop. Ext. MF3029. http://bookstore.ksre.ksu.edu/pubs/mf3029.pdf

SAS Institute. 2012. SAS/STAT user's guide, Version 9.3. SAS Inst., Cary, NC.

USDA-AMS. 2018. Hay Reports. Available at https://www.ams.usda.gov/market-news/hay-reports (verified 29 November 2021).

Table 1. Nitrogen fertilizer application rate effects on dry matter (DM) yield, N uptake, nitrogen use efficiency (NUE), nitrogen recovery efficiency (NRE), crude protein (CP) and nitrate (NO3) concentration over two growing seasons (2011 and 2012) near Randolph, KS.

N rate	Forage DM	N uptake	NUE	CP	1ST harvest NO_3	2ND harvest NO_3
	lb/acre		lb DM/lb	%	ppm	
0	4235 c[a]	46 c	91 b	7.1 c	420 b	23 a
25	4790 c	53 c	91 b	7.3 bc	351 b	32 a
50	5777 b	73 b	80 a	8.1 ab	887 b	60 a
75	6349 ab	82 ab	78 a	8.2 a	3181 a	78 a
100	6643 a	90 a	75 a	8.6 a	5484 a	143 a

[a]Means followed by same letter (s) within N rates are not significantly different using the least squares means (LSMEANS) multiple comparison procedure ($P > 0.05$).

Table 2. Nitrogen fertilizer application rate effects on dry matter (DM) yield, N uptake, nitrogen use efficiency (NUE), nitrogen recovery efficiency (NRE), crude protein (CP) and nitrate (NO3) concentration over four locations in western Kansas.

N rate	Forage DM	N uptake	NUE	CP	NO_3
	lb/acre		lb DM/lb	%	ppm
0	3854 c[a]	48 d	80 a	8.2 b	1675 b
25	4523 b	58 c	81 a	8.1 b	1953 b
50	5025 ab	66 bc	80 a	8.5 ab	3307 a
75	5366 a	69 ab	80 a	8.3 b	4005 a
100	5357 a	69 ab	76 a	8.7 ab	3741 a
125	5485 a	77 a	73 a	9.0 a	3551 a

[a]Means followed by same letter (s) within N rates are not significantly different using the least squares means (LSMEANS) multiple comparison procedure ($P > 0.05$).

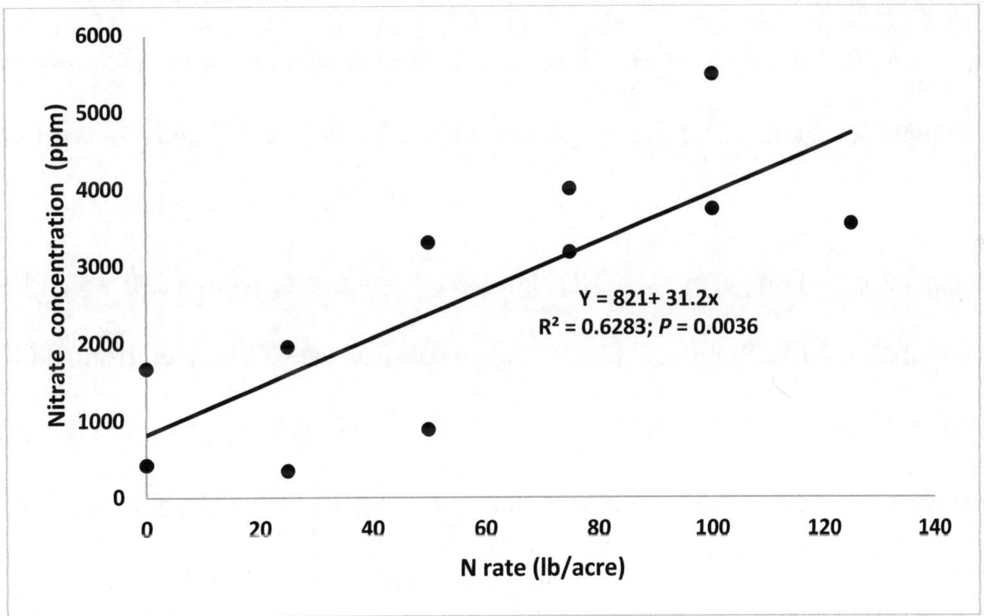

Fig. 1. Relationship between forage nitrate concentration and N fertilizer application rates. Data are averaged across six-site years.

UTILIZING COVER CROPS FOR FORAGE IN THE SEMI-ARID CENTRAL GREAT PLAINS

Augustine K. Obour[1], Johnathon D. Holman[2], Logan Simon[1], and Peyton Mauler[1]

Adoption of cover crops (CCs) is slow and not widely popular in the semi-arid central Great Plains (CGP) because CCs utilize water that otherwise would be available for the subsequent cash crop. Grazing or haying CCs can provide economic benefits to offset revenue loss associated with decreased crop yields following CCs. The study determined forage productivity and quality of CCs mixtures and quantified residue retention when CCs are utilized for forage in dryland cropping systems. Treatments were spring or summer-planted CCs grown in place of fallow, which were hayed, grazed by yearling heifers, or left standing. The study was conducted from 2015 to 2020 in a winter wheat (*Triticum aestivum* L.)–grain sorghum (*Sorghum bicolor* L.)–fallow cropping system with all crop phases present in each block and year of the study. Results showed forage mass varied from year-to-year highly dependent on growing season precipitation and temperature. Forage mass of spring-planted cover crops ranged from 3145 lb/acre in 2015 to 1655 lb/a in 2019. Similarly, forage mass of post-wheat CCs ranged from 1700 lb/acre in 2019 to 4600 lb/acre in 2018. Across the 6-yr study, crude protein concentration of spring-planted CCs averaged 12.6% compared with 8.8% for post-wheat CCs. On average, haying removed 62 to 73% of the aboveground CC biomass while grazing removed only 26 to 40%. Our findings suggest CCs can replace fallow to produce high-quality forage while improving soil health.

[1]Agricultural Research Center-Hays, Kansas State University, Hays, KS; [2]Southwest Research-Extension Center, Kansas State University, Garden City, KS.

OPTIMIZING TALL FESCUE HAYFIELDS THROUGH NITROGEN AND PHOSPHORUS FERTILIZATION

B.C. Pedreira[1], D.W. Sweeney, J.K. Farney, J.L. Moyer, and D.A. Ruiz Diaz1

Tall fescue [*Schedonorus arundinaceus* (Schreb.) Dumort.] is a cool-season, perennial grass grown on more than 1 million acres in Kansas. However, meadows and pastures on the claypan soils of the eastern Great Plains are still often under-fertilized. Our objective was to determine the effect of N and P fertilization management on the total annual forage accumulation (FA) of tall fescue 'Kentucky 31' in the early years of a stand. The experimental design was a split-plot arrangement of a randomized complete block with three replications on each of two adjacent sites established in sequential years. The four P_2O_5 rates of 0, 25, 50, and 100 lb/ac were the whole plots and the four N rates of 0, 50, 100, and 150 lb/ac were the subplots. Prior to each production year, P treatments (as triple super phosphate) were broadcast in the preceding fall and N treatments (as urea) were broadcast in late winter. For three years after the establishment year, the plots were cut in mid-/late-May and mid-/late-September and harvests were summed to obtain annual FA. Total FA differed depending on the phosphorus and nitrogen fertilization. At site 1, the total FA increased from 0 to 150 lb N/ac, at each of the P rates. However, the greater the P rate, the greater the total FA. At site 2, while the total FA was not affected by P fertilization rate, on average, for each lb of N applied, FA increased by 24.2 lb. Thus, N fertilization up to 150 lb/ac may increase tall fescue FA, and N fertilizer responses may be maximized when combined with P fertilization.

[1]Assistant Professor (pedreira@ksu.edu), Professor, Associate Professor, Professor Emeritus, Kansas State University, Southeast Research and Extension Center, Parsons, KS; and Professor, Kansas State University, Department of Agronomy, Manhattan, KS.

2020 RESULTS OF TIMED MECHANICAL WEED CONTROL IN EASTERN OHIO

T.G. Wiseman, C. Gelley, C.D., Penrose, J.S. McCutcheon, M.L. Landefeld[1]

Abstract

Maintaining a quality pasture is an essential part of all forage production systems because unwanted weeds compete for nutrients, water and sunlight needed for optimum growth. Forage growth in Ohio pastures is a critical part of farm production because grazing livestock are present on more than 36 percent of Ohio farms. Weeds can lower forage crop quality and productivity if left uncontrolled. Mechanically mowing pastures is one method producers can use if they do not want to spray herbicides. The purpose of this study is to determine if weed populations in pastured forages could be changed or reduced by varying the timing of mowing throughout the late spring and summer growing period without the use of herbicides. Treatments in this study consist of: (1) Control (no mowing), (2) June only mowing, (3) July only mowing, (4) August only mowing, (5) September only mowing, (6) June and August mowing, (7) July and September mowing, and (8) mown each month; June-July-August-September. The overall weed dry matter (DM) pounds/acre (lb/ac) shows the mowing in June or August were higher than all other treatments. The August treatment also yielded the most forages of 10437.0 lbs/ac on a DM basis. Mowing pastures in July and September resulted in the lowest quantity of forages of 4878.0 lbs/ac DM.

Introduction: Ohio farms have more than 1.28 million cattle and calves, more than 117,000 sheep and lambs and more than 1.42 million acres in pastureland (Turner and Morris, 2015). Topography, soil profiles and the existence of many woodland acres limit large scale agronomic row crop production. Most of the land in the Appalachian foothills of Ohio not in woodlands is used for production of cool season grasses for grazing livestock. Grazing livestock are used to harvest forages grown in pasture fields. Forages are an environmentally friendly crop that protects soils from erosion, improve soil tilth and enhance farm profitability. Weed growth in perennial forage pastures can reduce quality of the forage, quantity of useable forage and or cause livestock mortality if poisonous weeds are present and consumed. Some weeds are not palatable so livestock reject eating them. Some weeds produce chemicals that are harmful to other forage crop plants through a process called allelopathy and this can reduce growth of forage plants that are desired in the pasture. Mechanical control most often refers to mowing with a cycle bar mower or a rotary mower (brush hog). In combination with other control methods such as good fertilizer and liming practices and herbicides, mowing can be an effective tool in weed and brush management (Schnakenberg et al., 2010). The purpose of this study is to determine if weed populations in pastured forages could be changed or reduced by varying the timing of mowing throughout the late spring and summer growing period without the use of herbicides.

---

[1]Extension Educator, Ohio State University Extension; Extension Educator, Ohio State University Extension; Professor and Educator, Associate Professor and Assistant Director, Ohio State University Extension; Ohio State University Extension; retired Assistant Professor and Educator, Ohio State University Extension, respectively.

Broadleaf herbicides that control many problem weeds in Ohio pastures also reduce or eliminate legume plants that fix nitrogen from the air to feed grass forages. Adequate available nitrogen is critical for maximum grass growth in pastures. Jim Gerrish (2011) said "For an effective legume-based N supply, pastures should maintain a 30-50% legume component. Lower legume presence will lead to chronic N deficiency in the pasture and disappointing production.

Materials and Methods: This is a thirty-two (32) plot randomized complete block design with eight (8) treatments, including a control, and four (4) replications of each treatment. Each plot is fifteen feet wide by twenty feet long with an additional one foot border along each side to allow mechanical mowing equipment to be able to pass between marker posts. The site is a predominately fescue and mixed grass pasture field and the soil test (Mehlich III) results were taken in 2015 at the beginning of the multi-year experiment. Soil pH is 7.3, Phosphorus (P) 243 ppm, Potassium (K) 401 ppm, Magnesium (Mg) 456 ppm, Calcium 3037 ppm and the CEC 19.8(meq/100g). Forage samples were taken near the beginning of June, July, August and Septemberin 2015 and 2016. Each plot was rated for the amount of broadleaf weed pressure contained at the time of sampling. A Likert-Type scale of 0-9 was used (where 0 amounted to no visible weeds, up to 9 meaning 90% weeds). One additional rating was taken at the beginning of October before the end of the growing season.

Forage samples two feet by two feet ($4ft^2$) were hand harvested from each plot and broadleaf weed species were noted. Total weight of each sample was recorded, weed(s) removed if present and weighed, and the weed-free sample weightwas recorded. Any weed(s) removed from the samples were recorded by specie(s). From each ofthe 32 weed-free samples a sub-sample was removed to calculate dry matter. All samples were placed in the forage dryer at 46 degrees Celsius and remained there until dry. All weights were recorded in grams at that time and calculations were subsequently made from each sample to determine dry matter per acre. Each month, after harvesting samples, cow/calf pairs were used tograze the paddock, where the plots were laid out, until the desired amount of residual remained. Cattle were then removed until the next month. After each grazing, treatment plots were cut witha rotary mower according to the plan design, making one pass over the plot cutting to a height ofapproximately four inches above the soil surface. This was the fourth year of a five-year study. Treatments consist of: (1) Control (no mowing), (2) June only mowing, (3) July only mowing, (4) August only mowing, (5) September only mowing, (6) June and August mowing, (7) July and September mowing, and (8) mown each month; June-July-August-September.

Results: The study shows variation between plots when looking at the existing forage late in the growing season. Physical size of some weeds standout since they have not been mowed. Canada thistle and cocklebur are the two weeds most visible in the plots. However, there are a variety of other weeds such as: burdock, dandelion, horsenettle, ironweed, plantain and smartweed. While each growing season is different, rainfall amounts from the Eastern Agriculture Research Stationweather station, on location, are listed in Table 1.

Discussion: Quality forages in pastures is a key to preventing weed invasion and interference and optimizing animal performance. Weeds are opportunistic in their germination and establishment during adverse weather conditions. They favor open or disturbed areas which can be caused by livestock pugging soil when conditions in pastures are wet or over grazing. Perennial, biennial and annual broadleaf weeds can have a negative effect on livestock production, so some farm managers mow pastures to reduce targeted weeds. This on farm trial is an educational experiment to help landowners determine the best time, or times, to mow pastures if trying to reduce broadleaf weed pressure and not use herbicides.

Results from this study indicated mowing in June yielded the highest tonnage of forages, but also resulted in the most weeds as well. Mowing in June in most years would prevent grasses from producing seed heads and promote tillering. It may also be too early, opening the canopy increasing weed growth. Previous data would indicate that the June/August mowing dates may be best, but it is not significantly different than other mowing frequencies except the June only cutting (Landefeld, et al., 2017). If one can reduce weed pressure by mowing at the appropriate time, the need for herbicide applications may be reduced or eliminated. Also, when mowing multiple times a year, legume plants may thrive better and become a higher percentage of the sward over a period of time. These factors could have a positive effect on forage quality in the plots for the future. This trial will continue to determine if there will be significant differences in weed populations, and differences in the yield based on the timing of the mowing.

Conclusion: Mowing pastures once per year can be an effective in maximizing forage yield. The initial weed pressure may determine the number of times a pasture may need to be mowed. Regular mowing will also greatly reduce overall forage yield, but also had the lowest quantity ofweeds in this study. Regular mowing helps prevent weeds from establishing, spreading, and competing with desirable grasses and legumes (Curran and Lingenfelter, 2009). Mowing only in August produced the most forages, but this treatment also produced the second highest quantity of weeds.

Literature Cited:

Curran, W. & Lingenfelter, D. (2009). Weed Management in Pasture Systems. *Penn State University.* Available at: http://extension.psu.edu/pests/weeds/control/weed-management-in-pasture-systems/extension_publication_file

Gerrish, J. (2011). Managing Pasture Fertility, American Grazing Lands Services LLC, Available at: http://ucanr.edu/sites/placernevadasmallfarms/files/102317.pdf

Schnakenberg, T., Miller, P., Hobbs, J., Scheidt, J. & Kenyon, S. (2010). Pasture Weed and Brush Control. *University of Missouri.* Available at: http://extension.missouri.edu/stone/documents/ag/articles/PastureWeedBrushAug10.pdf

Turner, C. & Morris, B. (2015). Ohio Agriculture Statistics, 2015 Annual Bulletin. Available at: https://www.nass.usda.gov/Statistics_by_State/Ohio/Publications/Annual_Statistical_Bulletin/Master14.pdf

Landefeld, M., Penrose, C., McCutcheon, J., Wiseman, T. (2017). Timing Mechanical Weed Control in Pastures. Journal of the National Association of County Agricultural Agents. Available at: https://www.nacaa.com/journal/index.php?jid=775

Table 1. Rainfall measured in inches during the significant growth months

Year	May	June	July	August	September	5 Month Total
2020	5.1	3.4	2.3	3.7	2.6	17.1

Table 2. Monthly average weed weight

Treatments	Weed weights lbs/ac DM
Control	181.8
June	541.7
July	98.7
August	409.8
September	171.0
June and August	319.9
July and September	260.4
June, July, August and September	196.1

Table 3. Monthly average forage weight

Treatments	Forage weights lbs/ac DM
Control	7034.0[a]
June	5459.1[a]
July	7456.4[a]
August	10437.4[a,b]
September	7413.7[a]
June and August	7387.0[a]
July and September	4784.3[a,b]
June, July, August and September	6298.4[b]

MOTIVATIONS AND ISSUES FACED BY LIVESTOCK PRODUCERS PRESERVING FORAGES

T. Picado-Pérez[1], R. Lemus[2,3], D. Rivera[4], and L. Villalobos[5,6]

Anthropogenic activities have generated an accelerated and irreversible change in the climate, and livestock systems are being affected. Technologies that promote sustainability, increase resilience, and profitability are needed. Forage conservation can address dry matter shortages at critical times; planning to plant areas or utilizing forage surplus can sustain production while avoiding dependence on external inputs. Producers using preserved forages have the same objective, but systems, motivational factors, productive criteria, and limitations are diverse. To investigate this diversity, we surveyed 47 producers from different regions and livestock systems throughout Costa Rica. We found that 56.52% grow fodder for self-consumption and preserved forages are more frequently used in dairy than in beef or dual-purpose systems. Silage is the predominant technology and hay is the least. Producers used hay throughout the year in 63.16% of farms, haylage in 40% of farms, and silage in 43.75% of farms. Producers indicated that the main limiting factor for preserved forages was the high cost of inputs, followed by access to equipment and machinery and experience in preserving forages. However, we did not find that the use of preserved forages negatively or positively affects production levels. Our study shows that producers in Costa Rica are currently preserving forages to supplement their livestock during critical times, however, lack of training in these techniques may be leading to varied results that in turn could be impacting their successful implementation.

[1]Master Program in Animal Nutrition, Univ. of Costa Rica, [2]Center for Forage by Management & Environmental Stewardship, Mississippi State Univ., [3]Dept. of Plant and Soil Sciences, Mississippi State Univ., [4]Southwest Research & Ext. Center, University of Arkansas, [5]Research Center for Animal Nutrition, Univ. of Costa Rica, [6]Department of Animal Science, University of Costa Rica. Corresponding author: luis.villalobosvillalobos@ucr.ac.cr.

IMPACTS OF GRAZING COVER CROPS ON ANIMAL PERFORMANCE, CROP PRODUCTION, AND SOIL HEALTH IN EAST-CENTRAL MISSISSIPPI

J.B. Rushing[1], J.C. Lyles[2], B.S. Bass[3], and K.R. Waddell[3]

Integrated crop-livestock systems (ICLS) are designed to produce both cash crops and livestock simultaneously on the same land base, which can reduce nutrient concentration and environmental risk, improve soil structure and function, exploit natural pest control processes, and increase overall farm productivity and economic sustainability. A field trial was conducted at the Coastal Plain Branch Experiment Station in Newton, MS on a Prentiss fine sandy-loam soil during the 2019-2021 growing seasons. The trial consisted of two experimental areas (grazed and un-grazed) each with a randomized complete block design with three replications. Cover crop treatments included oats (O), O + crimson clover (OC), and OC + radish (OCR). Tillage treatments were superimposed on the un-grazed area (no-till and conventional till). This allowed for the following comparisons: 1. grazed, no-till (GNT); 2. un-grazed, no-till (UNT); 3. un-grazed, conventional till (UCT). The OCR treatment had significantly lower average daily gain (ADG; 3.03 lb hd^{-1} d^{-1}) from grazing beef steers than the other two treatments. Soybean yield was unaffected by cover crop species or tillage ($P > 0.05$). Tillage and grazing did not impact soil penetration ($P > 0.05$). Soil organic carbon (SOC) was significantly greater in GNT ($P < 0.0001$; 1.44%) compared to UNT and UCT. For GNT, net economic returns were lower for the OCR treatment (\$749.31 ac^{-1}) compared to the other two cover crop treatments.

[1]Associate Research/Extension Professor; Coastal Plain Branch Experiment Station, Newton, MS; [2]Research Associate I; Coastal Plain Branch Experiment Station, Newton, MS; [3]Graduate Research Assistant; Coastal Plain Branch Experiment Station; Newton, MS.

EFFECT OF SPECIES AND MATURITY ON YIELD, DIGESTIBILITY, AND PROTEIN CONTENT OF WINTER HARDY SMALL CEREAL GRAIN SILAGE

Abigail Sartin, Kallie Calus, Morgan Grabau, Alyssa Kuhn,
Daren Redfearn and Mary Drewnoski[1]

Winter-hardy small cereal grains can be planted in fall and harvested in the spring as silage, offering the potential for a double crop system with a warm-season forage or cash crop. The purpose of this study was to evaluate the silage yield and nutritive value of cereal rye, triticale, and wheat harvested at boot, pollination, milk, and soft dough. Dry matter yield of rye and triticale did not differ except at soft dough stage where triticale was greater than rye. Triticale yield was greater than wheat at pollination and soft dough, with rye being greater than wheat only at soft dough. In terms of energy content measured as digestible organic matter (DOM) after fermentation, rye and wheat did not differ and were both greater than triticale. Across all species, boot stage had the greatest DOM concentration, followed by pollination then soft dough with milk having the lowest energy content. Crude protein (CP) of all species decreased with increasing maturity. There were minor differences in CP among species, with rye being greater than triticale at boot, pollination, and soft dough but not differing from wheat. When energy and protein content were combined with dry matter yield and evaluated on a yield of DOM and CP per acre, rye and triticale had greater nutrient yields than wheat. It appears that triticale had slightly greater DM yield with slightly lower digestibility (energy) and protein content while rye had slightly lesser yields with slightly greater digestibility and protein content. For high quality forage, harvest at pollination appeared to have increased yield without sacrificing nutritive value. For maximized yield, harvesting at soft dough is a better option.

[1]University of Nebraska.

NOVEL PEST MANAGEMENT STRATEGIES TO CONTROL SMUTGRASS IN SOUTHERN FORAGE SYSTEMS

Nicholas J. Shay, Lisa L. Baxter, Nicholas T. Basinger, and Brian M. Schwartz[1]

Smutgrass (*Sporobolus indicus*) is a perennial, warm-season grass weed that continually invades bahiagrass (*Paspalum notatum*). A small plot trial was established to evaluate the effectiveness and economics of integrated weed management strategies on controlling smutgrass in bahiagrass grown in Alapaha, GA. A new location was defined annually in this two-year experiment. Plots were arranged in a three by four factorial arrangement with six replications per location. Herbicide treatments included Indaziflam (PRE), Hexazinone (POST), combination of pre + post-emergent herbicides (PRE + POST), and an unsprayed control. Fertilizer treatments included 50 units N/ac (N), 50 units N/ac + 50 lbs K_2O/ac (N+K), and an unfertilized control. Herbage accumulation was greater in plots receiving N fertilization at the second and fourth harvest events ($P < 0.01$). A decrease in herbage accumulation was noted for plots during the last two harvest events following the POST application ($P < 0.01$). Initially, all plots had a similar visual concentration of smutgrass ($P = 0.59$). However, at the end of the season less smutgrass was observed in the plots receiving POST ($P < 0.01$). Even though smutgrass declined, other annual or perennial weeds did not increase ($P = 0.08$). A more thorough statistical analysis is required, however the preliminary conclusion is that the combination of pre- and post- emergent herbicides with N and K fertilizer result in greater elimination of smutgrass and less bahiagrass injury than post-emergent herbicide alone.

[1]Graduate Student, University of Georgia, Crop and Soils Science Dept., University of Georgia, Tifton, GA (nicholas.shay@uga.edu); State Forage Extension Specialist, Crop and Soils Science Dept., University of Georgia, Tifton, GA (baxterl@uga.edu); Assistant Professor, Crop and Soils Science Dept., University of Georgia, Athens, GA (nicholas.basinger@uga.edu); Professor, Crop and Soils Science Dept., University of Georgia, Tifton, GA (tifturf@uga.edu).

POST-WHEAT SUMMER COVER CROP EFFECTS ON CROP YIELDS AND SOIL PROPERTIES IN A NO-TILL DRYLAND CROPPING SYSTEM

L.M. Simon[1], A.K. Obour[1], J.D. Holman[2], M.E. Schipanski[3],
S.K. Johnson[4], and K.L. Roozeboom[5]

Abstract

Traditional dryland cropping systems in the semi-arid Great Plains include long fallow periods of up to 14 months to conserve soil moisture. However, water storage in such systems are inefficient even under continuous no-till (NT). As less water is needed to produce forage compared to grain, cover crops (CCs) can be integrated for increased soil cover and potentially greater income when grazed or hayed as forages. An experiment was initiated in 2016 near Brownell, KS to investigate the effect of summer CCs grown after wheat on crop yields in a NT winter wheat (*Triticum aestivum* L.)-grain sorghum (*Sorghum bicolor* Moench.)-fallow cropping system. Cover crops were grazed, hayed, or left standing (no forage removal) and were compared to fallow with no CC. A second experiment was initiated in 2019 on cooperative producer fields near Hays and Alexander, KS to quantify the effect of grazing summer CCs on soil chemical and physical properties in NT dryland cropping systems. At these sites, grazed CCs were compared to un-grazed CCs (no forage removal). Across years, CCs produced 3446 lb/ac dry matter at Brownell. Haying removed 62% of the available forage mass while grazing removed only 39%. Grain sorghum yields were 11% less following standing and grazed CCs compared to fallow (68.6 bu/ac) but yields following hayed CCs were similar to fallow. Soil properties were unaffected by CC grazing compared to un-grazed CCs at Hays and Alexander. Our results showed post-wheat summer CCs can produce considerable forage mass with more residue retained following grazing compared to haying. However, grain sorghum yields following CCS could be reduced compared to fallow when all or most of the CC forage mass is retained, possibly due to nitrogen immobilization. Grazing CCs can compensate for reduced grain yields without negatively impacting soil properties.

Introduction: Integrating cover crops (CCs) to replace portions of the fallow periods in semi-arid dryland cropping systems has the potential to improve soil health by increasing soil carbon, reducing compaction, and enhancing soil structure. Previous research has emphasized the fallow period following summer crop [corn (*Zea mays* L.), grain sorghum (*Sorghum bicolor* Moench.), or proso millet (*Panicum milliaceum* L.)] harvest before winter wheat planting in the three-year winter wheat (*Triticum aestivum* L.)-summer crop-fallow cropping system of the semi-arid central Great Plains of the United States (Nielsen et al., 2015; 2017). However, less attention has

[1]Agricultural Research Center-Hays, Kansas State University, Hays, KS; [2]Southwest Research-Extension Center, Kansas State University, Garden City, KS; [3]Department of Soil and Crop Sciences, Colorado State University, Fort Collins, CO; [4]Northwest Research-Extension Center, Kansas State University, Colby, KS; [5]Department of Agronomy, Kansas State University, Manhattan, KS.

been paid to the fallow period following wheat harvest. As this region has a summer-dominant precipitation pattern, the months following wheat harvest has the highest likelihood for rainfall and could provide an opportunity for warm-season CCs to be incorporated (Mahama et al., 2016). Although the benefits that CCs provide for soil health have been recognized, their adoption in semi-arid dryland crop production remains limited due to subsequent grain yield penalties when soil moisture is reduced following CCs (Holman et al., 2018). However, utilizing CCs as an annual forage for livestock could offset losses in subsequent crop yield to balance goals of environmental and economic sustainability in dryland cropping. Still, concerns with grazing CCs include reduced soil organic carbon (SOC) accrual, increased soil compaction, and degraded soil structure compared to un-grazed CCs especially in NT production systems. The objectives of this study were to determine the forage accumulation of post-wheat summer CCs as well as the effects on grain sorghum yields and soil properties in NT dryland cropping systems.

Materials and Methods: The first experiment was conducted from 2016 to 2021 near Brownell, KS to investigate the effect of post-wheat summer CCs on crop yields in a NT wheat- sorghum-fallow cropping system. Cover crops were grazed, hayed, or left standing (no forage removal) and were compared to traditional fallow with no CC. A second experiment was conducted from 2019 to 2021 on two cooperative producer fields near Hays and Alexander, KS to further test the influence of grazing CCs on soils properties in NT dryland cropping systems. Annual precipitation is summer-dominant and averages 22 inches at these three sites.

The experiment at Brownell was arranged in a split-plot randomized complete block design. Main plots were the three crop phases of the wheat-sorghum-fallow crop rotation and split-plots (600 ft2) included three CC treatments and fallow. Cover crops were a mixture of forage sorghum, pearl millet (*Pennisetum glaucum* L.), sunn hemp (*Crotalaria juncea* L.), and cowpea (*Vigna unguiculata* L.) at seeding rates of 7.5, 2.5, 5, and 20 lb/a, respectively. Cover crops were planted into wheat stubble shortly after harvest (approximately July 1) and were either harvested for forage, mob-grazed, or left standing (no forage removal). Prior to grazing, available forage mass was determined from samples that were hand-clipped to ground level in two areas of 3 ft × 2 ft from each plot. Fresh weights were recorded and samples were oven-dried at 122°F for a minimum of 48 hours. Grazed CCs were stocked with yearling heifers (1500 lb/ac live weight) and moved daily for four days. Following grazing, CC residue retained was measured as previously described. Hayed CC treatments were harvested at 6 inches cutting height from a 3 ft × 100 ft strip in the middle of each plot using a Carter small-plot forage harvester (Carter Manufacturing Company, Brookston, IN). Whole plot weights were recorded with sub-samples collected and weighed. Sub-samples were oven-dried at 122°F to determine dry matter (DM) yield. No chemical termination was needed as CCs were selected to be terminated by frost (approximately October 15). Grain sorghum was planted the following June and harvested for grain in October using a Massey-Ferguson 8XP plot combine (Kincaid Equipment Manufacturing, Haven, KS).

Whole fields at Hays and Alexander were 50 and 80 acres, respectively. Four areas of each field were assigned as fenced zones to exclude grazing (un-grazed treatment) and cattle were allowed access the adjacent unfenced areas (grazed treatment). Cover crops were a mixture of forage sorghum, German millet (*Setaria italica* L.), sunflower (*Helianthus annuus* L.), sunn hemp, and radish (*Raphanus stivus* L.) at seeding rates of 7.5, 5, 1, 1, and 1 lb/ac, respectively. At Hays, CCs were grazed with cow-calf pairs (278 lbs/ac live weight) and, at Alexander, CCs were grazed with yearlings (459 lbs/ac live weight) for a target duration of 35 days. To determine the effect of CC management on soil properties, samples were collected in the spring before sorghum planting in 2020 at Hays and in 2021 at Alexander. Two intact soil cores of 6 inches in depth and 2 inches in diameter were randomly taken from each plot to determine soil bulk density. Samples were dried at 221°F for a minimum of 48 hours and bulk density was computed as mass of oven-dried soil divided by volume of the core. Ten additional 6-inch cores were collected randomly from each plot to determine SOC and nutrient concentrations. Additional soil samples were collected from the 0- to 2-in. soil depth with a flat shovel for the determination of water stable aggregates (WSA). Samples were gently passed through sieves with 0.32- to 0.19-in. mesh and allowed to fully air-dry. Two sub-samples from each replicate were used to estimate WSA by the wet-sieving method. Statistical analyses were completed using PROC GLIMMIX of SAS ver. 9.4 (SAS Institute, 2012, Cary, NC) with year and treatment considered fixed and replication considered random. Differences were considered significant at $P \leq 0.05$.

Results and Discussion: On average, post-wheat summer CCs produced 3446 lbs/ac DM at Brownell. However, CC productivity varied substantially from a high in 2018 to a low in 2019 (Figure 1a). On average, haying removed 62% of the available forage mass while grazing removed only 39%. Cover crops failed to emerge in 2017 due to extended summer drought conditions. Grain sorghum yields following CCs were 11% less with standing or grazed CCs compared to fallow (68.6 bu/ac) but yields following hayed CCs were similar to fallow (Figure 1b). Reductions in grain yield following CCs were observed in 2017 and 2020 when sorghum experienced summer drought conditions, but not in 2018 and 2019 when precipitation was favorable. At Hays, and Alexander, soil bulk density, SOC, NO_3, P, and mean weight diameter of WSA were unaffected by CC grazing compared to standing CCs (un-grazed) (Table 1).

In conclusion, post-wheat summer CCs can produce considerable forage mass with more residue retained with grazing compared to haying. However, subsequent grain sorghum yields could be reduced compared to fallow when all or most of the CC forage mass is retained (standing or grazed) but not when CCs are removed as hay. Soil chemical and physical properties when CCs are grazed on NT fields are similar to when CCs are left standing. These findings indicate that such CCs can be integrated after wheat in the NT cropping systems of the central Great Plains to supply forage for livestock in order to compensate for reduced subsequent grain yields without negatively impacting soil properties compared to standing CCs.

Literature Cited:

Holman, J.D., K. Arnet, J. Dille, S. Maxwell, A. Obour, T. Roberts, K. Roozeboom, and A. Schlegel. 2018. Can cover or forage crops replace fallow in the semiarid central Great Plains? Crop Sci. 58:932-944.

Mahama, G.Y., P.V.V. Prasad, K.L. Roozeboom, J.B. Nippert, and C. W. Rice. 2016a. Cover crops, fertilizer nitrogen rates, and economic returns of grain sorghum. Agron. J. 108:1-16.

Nielsen, D.C., D.J. Lyon, G.W. Hergert, R.K. Higgins, and J.D. Holman. 2015. Cover crop biomass production and water use in the central Great Plains. Agron. J. 107:2047-2058.

Nielsen, D.C., D.J. Lyon, and J.J. Miceli-Garcia. 2017. Replacing fallow with forage triticale in a dryland wheat-corn-fallow rotation may increase profitability. Field Crops Res. 203:227-237.

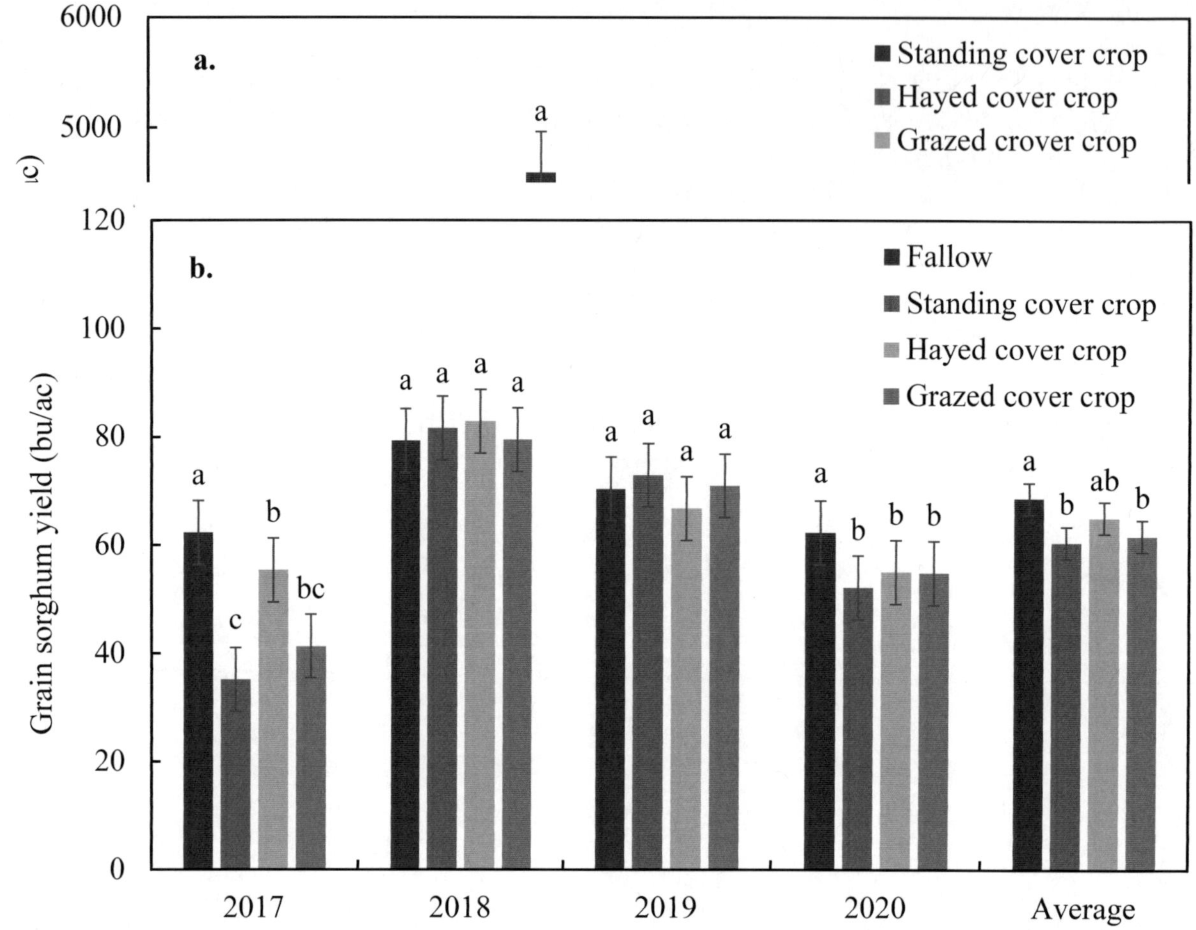

Figure 1. Cover crop residue retained following forage harvest (a.) and effect on subsequent grain sorghum yields (b.) near Brownell, KS.

[a]Error bars indicate standard error (α =0.05) and bars with the same letter are not significantly different (α =0.05) among treatments within the same year.

Table 1. Effects of post-wheat summer cover crop management on soil physical and chemical properties in the 0- to 2- and 2- to 6-inch soil depths.[a]

			BD[b]	SOC	NO_3-N	P	MWD
Location	Depth	Treatment	g cm^{-3}	lb/ton	ppm	ppm	inch
Hays	0- to 2-inch	Ungrazed	1.25	41.0	14.4	48.3	0.077
		Grazed	1.32	37.8	16.4	45.2	0.057
	2- to 6-inch	Ungrazed	1.38	30.6	5.2	24.6	-
		Grazed	1.41	31.6	8.7	23.8	-
Alexander	0- to 2-inch	Ungrazed	1.36	24.8	7.0	33.7	0.063
		Grazed	1.24	28.0	9.7	42.0	0.056
	2- to 6-inch	Ungrazed	1.46	18.0	3.5	15.2	-
		Grazed	1.41	18.6	3.6	8.27	-

[a]No significant differences at α=0.05.

[b]BD, bulk density; SOC, soil organic carbon; NO_3-N, nitrate-nitrogen; P, phosphorus; MWD, mean weight diameter of water stable aggregates.

EFFECTS OF THATCH DEPTH ON NATIVE WARM-SEASON GRASS (NWSG) ESTABLISHMENT

K.J. Swilling, E. Bisangwa, J. Mansfield, C. Bitler, P.D. Keyser[1]

A challenge to successful establishment of Native warm-season grasses (NSWGs) is that due to their small seed size, they need a high-quality seedbed. In addition to weed control, producers must ensure that heavy thatch is removed before planting. Thresholds for thatch left on the seedbed without interfering with establishment have not been documented. We evaluated the effects of thatch depth and composition on NSWG seedling recruitment. Treatments were imazapic (herbicide) vs. no imazapic, grazing vs. no grazing, and 0-, 1-, 3-cm (2019), 0-, 2.5-, 4-cm (2020) deep hay thatch. A blend of big bluestem (*Andropogon gerardii*), little bluestem (*Schizachyrium scoparium*), and indiangrass (*Sorghastrum nutans*) was drilled into the plots after placement of thatch. We recorded NWSG seedling numbers every three weeks post planting. Effects on NWSG populations were observed for thatch depth ($P < 0.01$), imazapic treatment × week ($P < 0.02$), and thatch depth × week ($P < 0.03$); grazing did not affect seedling density ($P < 0.50$). Thatch depth had a strong impact, with the greatest seedling density observed with 0.0 cm [(23.7 m^{-2} (2019), 3.0 m^{-2} (2020)] and the lowest in the 3.0-and 4.0-cm [6.0 m^{-2} (2019), (2020) 0.7 m^{-2}] thatch. Seedling density with and without imazapic was similar until week 12 at which point the imazapic treatment was greater (14.5 vs. 7.8 m^{-2} (2019) and 2.2 vs. 0.08 (2020)). The results indicate that thatch depth plays an important role in seedling recruitment; even 1.0 cm of thatch reduced mean seedling numbers by >50% compared to the 0.0 cm treatment.

[1]Graduate Research Assistant, UT, Knoxville, TN (kswillin@vols.utk.edu); Postdoctoral Researcher, UT, Knoxville, TN (ebisangwa@vols.utk.edu); Research Specialist, Greenacres Foundation, Cincinnati, OH (jmansfield@green-acres.org); Director of Research, Greenacres Foundation, Cincinnati, OH (cbitler@green-acres.org); Professor and Director, CNGM, UT, Knoxville, TN (pkeyser@utk.edu).

ELECTRIC FENCING FOR SERIOUS GRAZERS: A HANDS-ON PROGRAM FOR PRODUCERS AND AGRICULTURAL PROFESSIONALS

C.A. Thrailkill[1], C.D. Teutsch[2], M.D. Hayes[3], and J. McGill[4]

Electric fencing can be a highly effective tool for managing grazing when properly installed and maintained and an extremely frustrating experience when it is not. In 2020, the Kentucky Forage and Grassland Council and UK Master Grazer program partnered with Gallagher US to develop and implement a one-day hands-on workshop that focused on properly installing new electric fencing and troubleshooting existing electric fencing. The initial workshops were offered as in-service trainings to UK extension agents. Two classroom presentations on electric fencing "theory and principles" and "design and layout" were held in the morning. Following lunch, hands-on sessions were held in the field. These sessions included installation and troubleshooting energizers and grounding systems, utilizing and troubleshooting temporary fencing, using electrical offsets on new and existing fencing, and using fault finders to find and repair shorts. In a post workshop survey, 100% of respondents agreed or strongly that they had a better understanding of electric fencing theory and principles and how to design and install an effective electric fencing system. They also agreed or strongly agreed that they that they were better able to troubleshoot existing electric fencing. Written comments included "excellent program", "terrific program, I liked the relaxed setting and hands-on demos", and excellent speakers and presentations…well worth my time attending".

[1]Master Grazer Coordinator and [2]Extension Specialist (Chris.Teutsch@uky.edu), UK Research and Education Center, Princeton, KY; [3]Assistant Extension Professor, BAE, University of Kentucky, Lexington, KY; and [4]Gallagher, Riverside, MO.

TEPARY: A DROUGHT TOLERANT FORAGE FOR THE SOUTHERN GREAT PLAINS?

T.W. Witt, B. K. Northup, and C. A. Urrea Florez[1]

Tepary (*Phaseolus acutifolius* (A.) Gray is traditionally grown in the Sonoran Desert of the southern U.S. and northern Mexico as a grain crop. Ongoing and past research in Oklahoma has shown that tepary is a drought tolerant plant with biomass production and forage quality similar to soybean in rainfed conditions. In this study, ten tepary genotypes and one forage soybean (control) were grown at El Reno, OK to determine biomass, forage quality [acid detergent fiber (ADF), crude protein (CP), in vitro true digestibility (IVTD), neutral detergent fiber (NDF), percent carbon (C), percent nitrogen (N), and relative feed value (RFV)], cutting height, and cutting frequency required for tepary production. In 2020, tepary and soybean were planted at 21 seeds/ft^2 in 10 inch rows. At 30- or 45-day intervals, tepary and soybean plants were cut at 2, 4, or 6 inches above the soil. After harvesting, fresh and dry weights were taken and samples were ground to 0.08 inches using a Wiley mill. ADF and NDF were evaluated using the ANKOM method for a batch fiber analyzer. CP was calculated by multiplying %N by 6.25. IVTD was obtained by using the procedures for a Daisy incubator. The forage soybean plots were eaten by wildlife (possibly deer), so no measurements were observed. Dry matter production by tepary ranged from 2264 to 443 lb/ac when cut to a 2-inch height, 1209 to 391 lb/ac when cut to a 4-inch height, 1110 to 382 lb/ac when cut to a 6-inch height, and 17153 to 5455 when uncut. This data indicates that tepary may be harvested at multiple heights during growing seasons, but biomass yield will be greatly diminished.

[1]Research Agronomist and Research Ecologist, USDA-ARS, Grazinglands Research Laboratory, El Reno, OK (travis.witt@usda.gov, brian.northup@usda.gov); Dry Bean Breeding Specialist, UNL, Panhandle Research and Extension Center, Scottsbluff, NE (currea2@unl.edu).

AUTHOR INDEX

Aguerre, M. 7, 8
Aiken, R. 20
Anderson, B. 5
Assefa, Y. 24
Bagley, C. 2, 3
basinger, N. 63
Bass, B. 61
Baxter, L. 4, 5, 34, 63
Beckman, B. 33
Berry, M. 6
Biebler, C. 33
Billman, E. 1
Bisangwa, E. 6, 70
Bitler, C. 70
Boss, M. 7, 8
Brown, P. 34
Bush, J. 9, 10, 11, 14
Calus, K. 62
Campbell, B. 1
Catunda, K. 12
Churchill, A. 12, 13
Clark, K. 33
Diaz, D. 55
Dooley, B. 14
Drewnoski, M. 33, 62
Farney, J. 55
Fick, W. 15
Florez, C. 72
Gates, R. 34
Gelley, C. 56
Ghaste, M. 20
Goslee, S. 16
Gotsick, E. 17
Grabau, M. 62
Grev, A. 18, 19
Gruss, S. 20
Guretzky, J. 22
Han, K. 3
Harmoney, K. 21, 22
Hayes, M. 71
Hendrix, B. 14, 23
Henning, J. 9, 10, 11, 17
Holman, J. 20, 24, 37, 44, 49, 54, 64
Hudson, W. 5
Huff, J. 30
Ingram, S. 32
Isbell, F. 13
Jackson, J. 23
Johnson, A. 33
Johnson, K. 20
Johnson, S. 20, 37, 64
Keyser, P. 6, 42, 70
Kichler, J. 34
Knight, C. 34
Kuhn, A. 62
Laborie, E. 33
Lamm, F. 21
Landefeld, M. 56
Lea, K. 35
Lemus, R. 36, 60
Lesoing, G. 33
Lyles, J. 61
Mansfield, J. 70
Mauler, P. 24, 37, 44, 49, 54
McCutcheon, J. 56
McGill, J. 71
Mitchell, D. 42
Moore, B. 12
Morgan, S. 34
Morrow, A. 43
Moyer, J. 55
Northup, B. 72
O'Brien, D. 19
Obour, A. 20, 24, 37, 44, 49, 54, 64
Palfreyman, R. 2, 3
Pedreira, B. 55
Penrose, C. 56
Picardo-Perez, T. 60
Potts, S. 18
Power, S. 12
Ray, L. 34
Raymond, C. 23
Reay-Jones, F. 1
Redfearn, D. 33, 62
Rees, J. 33
Reyes-Esteves, R. 21
Rivera, D. 60
Roozeboom, K. 37, 64
Rushing, J. 61
Sapp, P. 34
Sartin, A. 62
Schick, B. 33
Schipanski, M. 64
Schoenian, S. 19
Schwartz, B. 63
Semier, J. 18

Semler, J. 19
Shay, N. 63
Sigua, G. 1
Simon, L. 24, 37, 44, 49, 54, 64
Smith, R. 6
Smith, S. 9, 10, 11, 17, 35
Strickland, J. 7, 8
Sweeney, D. 55
Swilling, K. 70
Tanner, S. 34
Teutsch, C. 9, 10, 11, 14, 17, 23, 71
Thrailkill, C. 71
Tucker, J. 4
Tuinstra, M. 20
Villalobos, L. 60
Waddell, K. 61
White, J. 36
Whitson, J. 2, 3
Widhalm, J. 20
Wiseman, T. 56
Witt, T. 72